D0950384

ALSO BY JOE AND TERESA GRAEDON

The People's Pharmacy-2 (Avon, 1980)

Joe Graedon's The New People's Pharmacy: Drug Breakthroughs of the '80s (Bantam, 1985)

The People's Pharmacy, Totally New and Revised (St. Martin's Press, 1985)

50+: The Graedons' People's Pharmacy for Older Adults (Bantam, 1988)

Graedons' Best Medicine: From Herbal Remedies to High-Tech Rx Breakthroughs (Bantam, 1991)

The Aspirin Handbook: A User's Guide to the Breakthrough Drug of the '90s (Bantam, 1993) with Tom Ferguson, M.D.

The People's Guide to Deadly Drug Interactions: How to Protect Yourself from Life-Threatening Drug/Drug, Drug/Food, Drug/Vitamin Combinations (St. Martin's Press, 1995)

The People's Pharmacy, Completely New and Revised (St. Martin's Press, 1996, 1998)

Dangerous Drug Interactions: How to Protect Yourself from Harmful Drug/Drug, Drug/Food, Drug/Vitamin Combinations (St. Martin's Press, 1999)

The People's Pharmacy Guide to Home and Herbal Remedies (St. Martin's Press, 1999)

Best Choices from The People's Pharmacy (Rodale, 2006, 2007)

Favorite Home Remedies from The People's Pharmacy (Graedon Enterprises, Inc., 2008).

Favorite Foods from The People's Pharmacy: Mother Nature's Medicine (Graedon Enterprises, Inc., 2009)

Recipes & Remedies from The People's Pharmacy (Graedon Enterprises, Inc., 2010)

ALSO BY JOE GRAEDON

The People's Pharmacy: A Guide to Prescription Drugs, Home Remedies, and Over-the-Counter Medications (St. Martin's Press, 1976)

No Deadly Drug (Pocket Books, 1992), a novel by Joe Graedon and Tom Ferguson, M.D.

ALSO BY TERESA GRAEDON

Chocolate without Guilt (Graedon Enterprises, 2002), a cookbook by Terry Graedon and Kit Gruelle

THE PEOPLE'S PHARMACY®

Quick & Handy HOME REMEDIES

THE PEOPLE'S PHARMACY®

Quick & Handy HOME REMEDIES

Q&As for Your Common Ailments

JOE *and* TERRY GRAEDON

NATIONAL GEOGRAPHIC
WASHINGTON, D.C.

IMPORTANT NOTE TO READERS

This book is not a substitute for the medical advice or care of a physician or other health care professional. The reader must consult a physician in matters relating to his or her health, especially with regard to any signs or symptoms that may require diagnosis or medical attention. Any health problems that do not get better promptly or get worse should be evaluated by an appropriate clinician and treated properly.

Home remedies are rarely tested in a scientific manner. They should not be considered a substitute for proper medical care. The reader should not assume that the home remedies discussed in this book are safe in all situations. Every treatment has the potential to cause some side effects for some people.

If you suspect that you or someone you care for is experiencing an adverse reaction from an herb, drug, dietary supplement, or home remedy, please consult a knowledgeable health professional immediately.

Throughout this book, the authors make references to specific products whose effects have been reported to them by readers, listeners, and website visitors. In publishing this book, the National Geographic Society does not endorse the use of any specific brands or products.

This book is dedicated to the thousands of readers and listeners who have contributed remedies, recipes, and commonsense suggestions to The People's Pharmacy over the past three decades. Without their input, this book would never have been possible.

We also dedicate it to the grandparents who have passed down their wisdom from generation to generation. It must not be forgotten.

CONTENTS

Introduction 10

12 **PART I. HOME REMEDIES**

Acne and Rosacea 15
Allergies 20
Anemia 22
Asthma 23
Back Pain 24
Body Odor 27
Brain Function 30
Bug Bites and Stings 32
Burns 36
Bursitis 39
Cancer 40
Canker Sores 43
Colds 48
Cold Sores 55
Colitis and Crohn's Disease 56
Constipation 59
Coughs 63
Cuts and Bruises 66
Dandruff 69
Diabetes 72
Diarrhea 78
Diverticulitis 84
Dry Skin 85
Eczema 88
Fibromyalgia 93
Gas 95
Gout 98
Headaches and Migraines 100
Heartburn 104
Hemorrhoids 116
Hiccups 117
High Blood Pressure 121
High Cholesterol and Triglycerides 128
Hot Flashes 135
Incontinence 140
Insomnia 142
Irritable Bowel Syndrome 144
Joint Pain and Arthritis 148
Kidney Stones 159
Lice 161
Macular Degeneration 163
Motion Sickness, Vertigo, and Dizziness 166
Muscle and Leg Cramps 170
Nail Fungus 177
Nausea 184
Nerve Pain 186
Nosebleed 190
Plantar Fasciitis 192
Psoriasis 193
Raynaud's Disease 199
Restless Leg Syndrome 200
Sex/Libido 205
Shingles 207
Sinusitis 209
Skin Fungus 210
Skin Tags 212
Stinky Feet 214
Sunburn and Sun Rash 217
Vaginal Dryness 218
Warts 220
Weight Loss 226
Wound Care 227

FAVORITE FOODS

Coffee 25
Blueberries 31
Green Tea 41
Pomegranate 42
Chicken Soup 52
Pineapple 57
Coconut 58
Cinnamon 74
Yogurt 83
Oolong Tea 92
Fennel Seed 96
Almonds 105
Broccoli 110
Hot Peppers 112
Beets 122
Chocolate 124
Fish, Fish Oil 131
Grape Juice 133
Walnuts 136
Cherries 153
Mustard 178
Ginger 188
Curry 198
Garlic 223

230 PART II. EATING FOR HEALTH

The DASH Diet 232
The Mediterranean Diet 235
The Low-Carb Diet 238

BREAKFAST
Anti-Inflammatory Curcumin Scramble 241
Joe's Brain-Boosting Smoothie 241
Cholesterol-Combating Oatmeal 242

LUNCH
Crustless Cauliflower and Red Pepper Quiche 243
Sole with Lemongrass Red Bell Pepper Salad 244
Wheat Berry Salad 244

DINNER
Cardamom Grilled Chicken with Mango Lime Sauce 245
Lentil Nut Loaf with Red Pepper Sauce 246
Pescado Al Cilantro 247

SIDES
Coleslaw with Mint 248
Farmers Market Saag 248
Roasted Garlic 249
Simple Salad Dressing 249

Sources 250
About the Authors 254
Acknowledgments 255
Recipe Contributors 256
Index 257

Introduction

TRADITIONAL ADVICE FROM AROUND THE WORLD is remarkably similar in its commonsense approach to staying healthy: Get plenty of exercise, sleep, and good food. Over the past 15 or 20 years, researchers have discovered that many ideas people once dismissed as "old wives' tales" have solid scientific underpinnings. Home remedies like cranberry juice for urinary tract infections or vitamin-D-rich cod-liver oil as a winter tonic have been proven helpful.

Grandmothers' wisdom fell out of favor with the medical establishment decades ago, however, and even the public has forgotten much of it. In keeping with a typically American fixation on all things instant, people have come to expect a pharmaceutical fix for their problems. Some doctors believe, with reason, that their patients would be disappointed to leave the office without a prescription.

Prescription drugs can be lifesaving, and our intention certainly isn't to disparage pharmaceuticals or the doctors who prescribe them. Millions of Americans depend on medications, and no one should ever change or discontinue a treatment regimen without the guidance of a doctor.

What concerns us, and the reason we conceived this book, is that we constantly hear from people who feel helpless and hopeless about caring for themselves and preventing the kinds of chronic conditions for which medication is required. As a culture, we've largely lost sight of how to live and eat so we can be healthy.

People have probably been using food as medicine for millennia. Studies have shown that other animals, too, seek out foods that may function as laxatives, antibiotics, or antidotes to toxins.[1] For thousands of years, people in regions as diverse as China, Japan, India, and ancient Greece have made use of food for its healing properties. The Greek physician Hippocrates, who lived 2,500 years ago, once famously proclaimed, "Let food be your medicine." (Hippocrates is also famous as the father of medicine and the originator of the Hippocratic Oath.) In recent years, as more researchers have become interested in the potential health benefits of various foods, scientific evidence that supports a food-as-medicine approach has begun to mount.

When we talk about using food as medicine, we mean two distinct things. First, we mean that the foods we choose to consume regularly can have a major impact on our health, and that even small changes to our diets may have significant long-term benefits, including the prevention of many chronic conditions. More intriguing is the possibility that certain foods might have specific therapeutic activities. People in India have been using turmeric, the yellow spice in curry and yellow mustard, for thousands of years. It adds a very special taste to foods, but it also fights inflammation. A surprising amount of research on curcumin, the active ingredient in turmeric, suggests great promise for using the spice to treat cancer, arthritis, and heart failure.[2] Studies indicate that beets are another example of a delicious food that can help control a common problem: high blood pressure.[3] Studies show that walnuts can lower cholesterol, help prevent heart disease,[4] and perhaps even help cut one's risk of developing type 2 diabetes.[5] Researchers have also confirmed that blueberries and other antioxidant-rich foods may help boost cognitive function in older people.[6] And the list goes on. This book contains dozens of other examples of literally using food as medicine to treat common ailments.

We also like to think of foods as home remedies. We've been collecting home and herbal remedies for decades—ever since the People's Pharmacy came into existence in 1976. Whenever we hear of a new remedy, we take note. If we hear the same remedy repeatedly, we like to share it. Our motto is, if it won't hurt, might help, and doesn't cost too much, it's worth considering. While not all remedies work for all people, not all pharmaceuticals do either.

Home remedies should never become substitutes for medical treatment, though, and you should always consult your physician before adding anything new to your regime, since some foods, herbs, and supplements may interact with medication. If you have an adverse reaction to a remedy, you should stop using it right away and talk to your doctor. But it may be worthwhile to try using foods or other remedies to combat some common ailments before going to the pharmacy.

Throughout these pages, you'll find information and citations from the medical literature on the medicinal properties of various foods. You'll also find recipes for both foods and remedies. We've invited a few of our favorite guests from the People's Pharmacy radio show—some of the country's leading experts on health and nutrition—to send us their favorite recipes for healthful eating and living. We've then selected from that already select group.

These experts' approaches to diet represent a diversity of opinions and approaches, but they're all based on the same basic principles of good food for good health. If anyone knows food, these folks do. We love food, too, and we've also contributed some of our own favorite recipes. We hope you enjoy them, and we hope they help you enjoy good health!

HOME REMEDIES

HOME REMEDIES REPRESENT a practical, affordable way to deal with some common ailments that don't necessarily need immediate medical intervention. Likewise, food chosen wisely can serve as Mother Nature's medicine. Hippocrates, the father of medicine, said, "Let food be your medicine." That's why this book has two parts. In the first section you will find some of our favorite home remedies that readers have been sharing with us for decades. In the second part, we discuss foods that can be used to promote good health. There is some overlap between home remedies and healing foods, and we trust you will find the information you need.

Remedies & Ailments

In the first part of the book you'll find 62 of the most common ailments that bedevil people, listed alphabetically, from Acne to Wounds. There's a wide range covered in between, both in terms of seriousness (from Colds to Crohn's Disease) and duration (from Bug Bites to High Blood Pressure). Within each category, we offer home remedies to help with these ailments.

Q A The ailments are presented in a question-and-answer format gleaned from our newspaper column, our radio show, and our website. Much of what we know about home remedies comes as a result of suggestions or queries from our readers and listeners. In a surprising number of cases, though, there is some scientific support for our approaches. In those cases, we offer the publication and the year in which the research was published. In many other instances these folk remedies have never been studied, and we cannot offer a scientific explanation for why they might work. We try to follow the motto "If it might help, won't hurt, and doesn't cost too much, it's worth considering."

Do check the Important Note to Readers on page 6 for the caveats one should always follow when experimenting with a home remedy.

Favorite Foods

Here and there throughout Part I, you will find that we focus on one of our 24 favorite foods. These have been found especially beneficial in preventing illness and maintaining health.

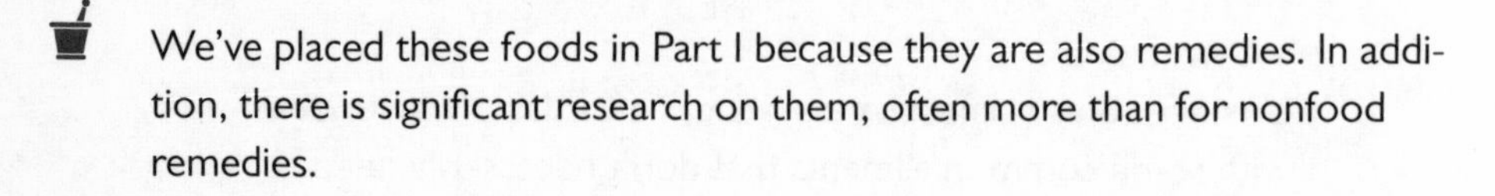

We've placed these foods in Part I because they are also remedies. In addition, there is significant research on them, often more than for nonfood remedies.

We also share recipes throughout the first part of the book, not only for remedies such as teas and cough syrups but also for delicious and healthful dishes that we have learned about from some of our favorite radio show guests or have invented in our own kitchen.

The two parts of our book really reinforce each other. Part II, Eating for Health, goes more deeply into healthful food choices, describing three diets with proven health benefits. There you will also find 13 additional recipes. You might turn straight to this section if eating well is your immediate priority.

We hope you find the remedies and recipes in this book tasty and helpful.

Acne and Rosacea

Acne is the bane of many teenagers, but it can also affect adults. These bumps and blemishes may be partly due to inflammation, triggered by certain skin bacteria. Ordinary acne (*acne vulgaris*) is a completely different condition from acne rosacea, although in the initial stages they may look similar. Rosacea affects women more than men and usually strikes during middle age. It causes redness of the cheeks, nose, and forehead. Dermatologists are still debating the causes of rosacea. Research suggests that one important factor is inflammation triggered by cathelicidin, one of the skin's innate immune defenses against bacteria, fungi, and some viruses.[1] Treatment for rosacea has involved oral antibiotics (doxycycline, minocycline, tetracycline) and topical antimicrobials (metronidazole). A topical gel containing azelaic acid (Finacea) can reduce the production of cathelicidin and improve symptoms. But the condition may not respond well to prescribed medications, leading to frustration. Home remedies are unproven but may be worth a try.

CORNSTARCH

Q Two years ago a dermatologist diagnosed my skin condition as rosacea and prescribed topical tetracycline and MetroLotion to be applied twice daily. My condition did not improve. The redness and rash were chronic and seemed to be getting worse. I tried all kinds of products, including over-the-counter lotions and cortisone creams. Then I put Argo Corn Starch on the rosacea. One place on my cheek near my nose looked especially bad. To my surprise, in a week it healed. Now all I do is wash my face morning and night, then apply a light coat of cornstarch. I have not had a recurrence of rosacea. My face is smooth and clear. Am I an isolated case, or is cornstarch a reliable treatment?

A Rosacea is a chronic skin condition that affects the chin, cheeks, nose, or central forehead. Redness, bumps, pimples, and visible blood vessels are common. Its cause is somewhat mysterious, but dermatologists often treat rosacea with oral antibiotics or topical anti-infectives like MetroGel or MetroLotion (metronidazole). Cortisone creams can make rosacea worse. Gentle face washing twice a day is recommended, but as far as we can tell, your cornstarch approach is unorthodox. We do not know if it would help anyone else or if you are an isolated case. The condition can wax and wane, but if this low-tech treatment works, count yourself fortunate.

DIET

Q I had facial acne and rosacea for at least four years. I blamed daily medications. A visit to the dermatologist did not help. Then one of your columns mentioned artificial sweeteners as a cause of diarrhea, so I stopped using them for that reason. When I quit drinking diet soda, my skin improved. Now, after six weeks, my skin problems are almost gone—for the first time in four years. I have diabetes and now drink only water or unsweetened drinks. Thank you.

A Dermatologists recognize that individuals have different triggers. We're glad to have helped you find yours. Doctors often prescribe oral antibiotics or topical antimicrobials. Some studies suggest that a topical B vitamin, nicotinamide (Nicomide), may also help control redness and bumps. Topical low-dose doxycycline (Oracea) is prescribed to maximize anti-inflammatory activity and to minimize antibiotic action.

Q My rosacea gives me a red nose and cheeks. Tetracycline helps but upsets my stomach. If I stop the antibiotic, the redness returns. My neighbor suggested two tablespoons daily of salsa. I started the salsa a month ago. Now my nose is not red or itchy as it usually is. It's hard to believe something that tastes so good could be good medicine. Have you ever heard of this treatment?

A Your experience with salsa is intriguing. Doctors usually tell rosacea patients to avoid food or drink that can dilate blood vessels, including hot beverages, alcohol, and spicy foods. We have heard of people using hot salsa for skin problems such as psoriasis and eczema, but this is the first time anyone has suggested it for rosacea. Capsaicin, the component in hot peppers that gives them a zing, has been tested topically for other skin conditions. Please let us know if it continues to work.

Q I read your article on acne disappearing when the writer gave up bananas. I too had this problem years ago. When I stopped eating bananas, my acne disappeared.

A We don't have a clue why some people might react to bananas by developing skin blemishes. While cutting out bananas might not work for others, it seems a simple experiment to try.

Q My 14-year-old daughter has had moderate acne for two years. Clearasil leaves bleach stains on her clothes. Antibiotics seem to make matters worse. The doctor suggested birth control pills, but it's not an option we'd entertain. Are there any topical or natural remedies that might work? What about diet?

A Any link between acne and diet is controversial. People once told teens to avoid chocolate and high-fat foods. That turned out to be unhelpful. However, research suggests that diet may make a difference.[2] Populations that eat low-carbohydrate diets that don't cause a rapid rise in blood sugar may be less prone to blemishes. Your daughter might try avoiding foods like candy, cookies, french fries, potato chips, sugar, and white flour to see if it helps her complexion. Ask your pharmacist about a topical treatment that contains a B vitamin. Nicomide-T Gel was equivalent to the topical antibiotic clindamycin in one controlled study.[3] Oral nicotinamide may also be helpful.[4]

LISTERINE

Q I have sensitive skin that reacts badly to everything. I have used Listerine for years to clear up small blemishes. Apply a dab to the area at night, and usually by morning the spot is clear. It doesn't irritate the surrounding skin either. My husband has started using it for shaving bumps too.

A **The herbal extracts and alcohol in Listerine that are supposed to "kill germs by millions on contact" may be useful in helping your blemishes heal. We have heard from other readers who have used Listerine in this way.**

MILK OF MAGNESIA

Q Have you heard of using milk of magnesia on severe acne? My son has cystic nodular acne. He is 16 and has been under a dermatologist's care for years. We have spent thousands of dollars to no avail. Recently he tried applying milk of magnesia to his face at night before bed. He looks better than he has in four years. Can you tell us why this is working?

A **Milk of magnesia is a solution of magnesium hydroxide and is best known for its laxative action. We don't know why it might combat acne, but we have heard that this laxative can help clear up seborrheic dermatitis.**

Q I am 44 years old and have had acne since my teens. Dermatologists have prescribed countless antibiotics, including Cleocin T, to no avail. Birth control pills worked, but when I stopped taking them, the acne returned. I also took Retin-A, which helped but caused sun sensitivity, redness, and cracking. I was excited to read about milk of magnesia as a topical treatment. My 12-year-old son and I are getting good results. Can milk of magnesia make acne disappear?

A **There are no good studies, though a letter in *Archives of Dermatology* suggests that a topical application of milk of magnesia each night**

could help reduce the redness and inflammation that is associated with acne. Some of our other readers have reported success with this remedy.[5]

RED CLOVER SALVE

Q I have been successful in keeping rosacea outbreaks from recurring by using a product from J. R. Watkins called Red Clover Salve. I simply rub a small amount on my nose and cheeks every morning. I find the salve just as effective as a prescribed ointment and it is also less expensive.

A A chronic skin condition, rosacea causes flushing and pimple-like outbreaks. The salve can be ordered online from Amazon.com. We found no studies recommending its use for rosacea, but we are glad it is helping.

VINEGAR

Q I have been suffering from rosacea for years. A dermatologist prescribed both metronidazole cream and minocycline twice daily. These were ineffective. I am a 47-year-old male, 5 feet 8 inches tall, and weigh 139 pounds. I exercise regularly (run and bicycle) and have a healthy diet. I drink alcohol occasionally, mainly red wine and beer. My cholesterol is low, and I take no medications. What else could I do for my rosacea?

A Alcohol is frequently blamed as a trigger for rosacea, so cutting back on beer and wine might help. Another nonstandard approach is antibacterial soap. One reader applies organic raw apple cider vinegar to the affected skin and washes it off with a gentle cleanser after 30 minutes. Another reports, "I have had rosacea—dry, flaky, reddened facial skin—for years. I decided to try vinegar as a facial cleanser. I dampen a cloth with it and wipe my face off once daily. My face has not felt this smooth or been this free of redness for a long time."

Allergies

Allergies cause a lot of misery, particularly at certain times of the year when pollen is in the air. Nasal congestion and sneezing are often accompanied by fatigue, mental fuzziness, and delayed reaction time. These symptoms can make driving hazardous, but so can many common antihistamines available over the counter to treat allergy symptoms. No wonder people get excited about home remedies to treat allergies—or to prevent them in the first place!

GLUTEN-FREE DIET

Q A caller on your radio show said that her allergies went away when she maintained a gluten-free diet. I have had a similar experience. I had battled allergies most of my life. About ten years ago I was diagnosed with lupus. Since eating a gluten-free diet for the last two years, I have been allergy free. In the last nine months I have had no lupus symptoms and have eliminated prescription medications. Gluten is toxic for me, and I will avoid it for the rest of my life. My teenage son was just diagnosed with celiac disease. I am hoping that a new gluten-free diet will eliminate his allergies as it did mine.

A Celiac disease is an inability to tolerate gluten, a protein found in barley, wheat, and rye. Celiac disease is very serious. It is not usually linked to allergies but may be associated with lupus.

NETI POT

Q Since using a neti pot daily, my friend, my daughter-in-law, and I have stopped our prescription nasal sprays and inhalers for sinus problems and allergies! I get bronchitis easily, but I have been cough free for two months since I began rinsing my sinuses nightly with a mixture of one cup of warm water and one-quarter teaspoon of plain salt.

A The neti pot looks a bit like Aladdin's lamp. It is a traditional technology for nasal irrigation to cleanse the nasal passages and sinuses. Practitioners of India's traditional ayurvedic medicine have used neti pots for regular nasal cleansing for hundreds of years. In using a neti pot, the head is tipped forward and slightly to the side so that water can be poured into one nostril and allowed to run out the other. Nasal irrigation may also be accomplished with spray equipment from a drugstore. A study suggests that many people with chronic sinus symptoms benefit from daily nasal irrigation.[1]

ROOIBOS TEA

Q While in Africa I started drinking rooibos tea every day. Now I am back home in Houston, and my usual fall allergies have not recurred. Have you heard of using rooibos tea for allergies?

A We have heard of rooibos (red bush) tea from South Africa. Traditionally it has been used to fight pollen allergies, but there is little clinical research to support its effectiveness. However, one study did suggest rooibos tea has an effect on the immune system that might help relieve allergic symptoms.[2]

STINGING NETTLE

Q A friend found a mention of nettle leaf for allergy relief in your book and passed it on. It works. When a student in one of my college classes told me that his allergy disrupted his sleep. I gave him a dose of my nettle leaf extract. An hour later he interrupted class to say his symptoms were gone. On your Web pages you discuss nettle root for prostate health. Are the uses of the leaf and the root different?

A Stinging nettle *(Urtica dioica)* is commonly used in Europe, both as medication and vegetable. In the United States, few know about it. You are correct that nettle root extract is used to treat symptoms of enlarged prostate. Research suggests that aboveground parts are useful in treating allergy symptoms.[3]

Anemia

Anemia is a relatively common condition, particularly in menstruating women and in people who follow a vegetarian diet. Nevertheless, anemia is a blood disorder and can be quite serious. It's caused by a lack of healthy red blood cells, which may be due to excessive blood loss or inadequate nutrition. Symptoms can include easy bruising, tiredness and lethargy, paleness, and lack of concentration. Pica—a craving for substances other than food (like ice, cornstarch, laundry starch, or dirt)—can also be a sign of anemia.

Anemia can sometimes signal an underlying condition, such as celiac disease or hypothyroidism. If you suspect that you suffer from anemia, consult your doctor. Iron supplements can frequently correct less acute anemia. Dietary alterations also may make a difference. Consider cutting out items such as wine, tea, and soy, which can block iron absorption. It also may be helpful to add iron-rich foods, such as spinach, beets, beans, and nuts, to your diet. For nonvegetarians, poultry, fish, or meat (especially liver) can provide iron from the blood protein hemoglobin, which the body more readily absorbs. Other foods rich in nonhemoglobin iron include quinoa (a grain), raisins, and blackstrap molasses.

BLACKSTRAP MOLASSES

You recently answered a question from a vegetarian blood donor who has low hemoglobin. He was concerned about caffeine. I too am a vegetarian and donate blood every 56 days. I do not consume caffeine, but my iron level at times has been too low to allow me to donate. I was told that tea (even herbal and decaf) robs the body of iron. So a week before I donate, I stop drinking tea. Since I started doing that, I have not had a problem with my iron level. For a hot drink before donating, the donor should try a tablespoonful of blackstrap molasses in hot water. It'll warm him up and provide iron.

Thanks for the recommendation for the iron-rich hot drink using blackstrap molasses. Caffeine doesn't affect iron levels, but many kinds of hot drinks have tannins and polyphenols that can interfere with iron absorption. Tea is rich in these compounds, and coffee and cocoa can also hinder iron absorption. So can herbal teas made from peppermint or chamomile.

Asthma

Asthma does not have a do-it-yourself treatment program. Even mild asthma requires medical supervision, and more severe asthma requires a thoughtful treatment plan. Doctors often give asthma patients a steroid inhaler and a bronchodilator to help open constricted airways. Combination products such as Advair have become popular, since they contain two drugs in one puffer. We're not here to discuss the pros and cons of asthma medications, but we can try to provide backup in a pinch. Imagine that you have packed your asthma medicine in carry-on luggage for a flight from Akron to Philadelphia. As you approach the boarding gate, however, you learn that the "puddle jumper" airplane cannot accommodate your overstuffed carry-on. The gate attendant unceremoniously tags your luggage, and off it goes to the underbelly of the airplane. Midway to Philly, your lungs get a bit twitchy and you start to wheeze. You can't ask the flight attendant to get your luggage. But you can ask for strong coffee, a remedy that often helps asthma sufferers.

CAFFEINE

Q I was wondering about coffee and asthma. Coffee works for my asthma! Which works better? Caf or decaf? Should it be brewed, or can it be instant? I am going on vacation this month, and it would make me feel better to know these things, in case I run into trouble.

A Physicians have known about the beneficial effects of coffee for treating asthma since at least 1859, when the Scots noted it in the *Edinburgh Medical Journal*. Research has shown that caffeine can open airways and improve asthma symptoms.[1] Caffeine is related to theophylline, an old-fashioned drug used to treat respiratory diseases such as asthma and emphysema. The recommended dose is usually three cups of strong coffee for an average adult. Decaf coffee will not work. Instant coffee contains less caffeine than brewed coffee, so a person seeking relief might need a few more cups of instant. Green tea also helps open airways but takes longer to do so. School nurses even use Mountain Dew soda for kids who left their inhalers at home. However, no one should rely upon coffee, green tea, or Mountain Dew to control asthma symptoms. Although these drinks can help in a pinch, prescribed medication offers more reliable relief.

Back Pain

Back pain plagues humankind. Almost everyone experiences a twinge now and again. Others are laid low for weeks, barely able to walk or to ride in a car without excruciating pain. Believe it or not, physicians still do not really know what causes acute low back pain. And there is a great deal of controversy about the best way to deal with this problem. A few decades ago, health professionals told people to rest and avoid moving for days or even weeks. Today, unless there is a herniated disc or other anatomical complication involved, doctors usually encourage patients to move gently as soon

THE PEOPLE'S PHARMACY

Favorite Food #1: Coffee

Bet you never thought your morning cup of joe might help reduce the risk of developing type 2 diabetes, cancer, stroke, heart disease, and even Alzheimer's. Coffee gets blamed for all sorts of things, from discolored teeth to headaches from caffeine withdrawal—it is habit-forming, after all. Caffeine also has been linked to increased risk for miscarriage in pregnant women,[1] and to breast cancer,[2] although the data remain mixed. Pregnant women certainly should be cautious about caffeine consumption. But for most people, the benefits of drinking too much coffee far outweigh the risks.

The news about this tasty beverage keeps getting better and better. Several studies suggest that coffee drinkers are at a lower risk for type 2 diabetes.[3] Surprisingly, coffee contains more antioxidants than many vegetables, and this may account for its cancer-fighting properties. In numerous studies, coffee has been associated with a decreased risk for cancers of the mouth, throat, esophagus,[4] colon,[5] kidneys,[6] liver,[7] skin,[8] and endometrium.[9]

Drinking coffee appears to help cut the chance of stroke as well. In one study, women drinking at least four cups of coffee a day seemed moderately less likely to suffer strokes than those consuming less than one cup a month.[10] Coffee also may help protect against heart disease, but these data are contradictory—maybe because unfiltered coffee might actually raise cholesterol. (If you have high cholesterol, you might consider trading in your French press or espresso maker for a drip-style brewer that takes a paper filter.)

Yet another boon is that coffee seems to lower the risk of Alzheimer's disease. Researchers analyzing the coffee-drinking habits of Finns in the 1970s and 1980s discovered that those who drank three to five cups daily were 65 percent less likely to be diagnosed with Alzheimer's several decades down the road.[11] Maybe it's time to wake up and smell the coffee.

as possible after a back spasm wears off. Some back experts criticize practitioners who perform spinal manipulation. Others think manipulation is about the only approach that makes any sense. We refuse to enter the fray. But we do offer some home remedies that are less likely to cause complications than more invasive approaches (such as back surgery). Of course, if acute lower back pain is accompanied by other symptoms, such as urinary control issues or numbness and tingling, we encourage a workup by a qualified back specialist.

MEDITERRANEAN DIET

Q I've suffered from severe lower back pain since 1964, when I was 14. In 1972, I was diagnosed with spondylolisthesis, fourth degree. In 1980 and 1986 I had surgical procedures to fuse my spine into place. In 2004, I was told my spine had slipped again; I had scoliosis and osteoarthritis in the spine and pelvis. Then, last year, I read your book *Best Choices from the People's Pharmacy* and began the Mediterranean diet at the beginning of September. I stopped taking Celebrex at that time. After three weeks, I was able to move my shoulders freely, despite previous bursitis, and no longer had pain in my spine and pelvis. When I stopped taking Lipitor, the muscle and nerve pains and the tingling in my extremities disappeared. I've been able to stop taking all medication for any sort of pain since December. Besides the freedom from pain, I love the feeling of well-being I've gained. I've shared the information with my cousins and friends and have given the title of your book to fellow patients I've met in the pharmacy or at clinics at the Princess Margaret Hospital here in Nassau.

A We are delighted that the anti-inflammatory benefits of the Mediterranean diet have relieved the pain from your serious back condition. This diet can also help prevent heart disease.

TURMERIC

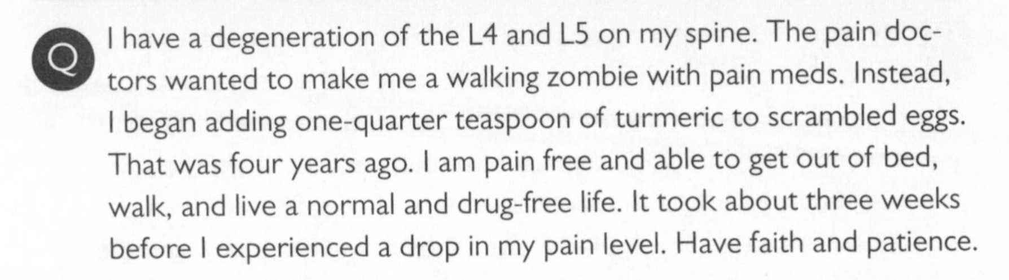

Q I have a degeneration of the L4 and L5 on my spine. The pain doctors wanted to make me a walking zombie with pain meds. Instead, I began adding one-quarter teaspoon of turmeric to scrambled eggs. That was four years ago. I am pain free and able to get out of bed, walk, and live a normal and drug-free life. It took about three weeks before I experienced a drop in my pain level. Have faith and patience.

A Thanks so much for sharing your experience and remedy. Following your suggestion, we found scrambled eggs a great way to get a dose of turmeric.

Body Odor

This condition can be extremely embarrassing, but in most cases doctors pay little attention to body odor because it isn't really a medical problem. Antiperspirants containing aluminum may help, but some people react badly to them, and many would prefer to avoid aluminum. Body odor may be caused by skin bacteria that break sweat down into volatile (smelly) compounds. Some people have found intriguing approaches to this problem.

BAKING SODA

Q Baking soda is fantastic for sweaty underarms. I've used it for many years because regular deodorants either cause an allergic reaction or don't work. Baking soda beats them all.

A Others have also found baking soda helpful. It is applied with a cotton ball.

LISTERINE

Q My daughter is entering puberty and dealing with underarm body odor. We tried many different deodorants and antiperspirants to no avail. I figured if Listerine kills the germs that cause bad breath, it might kill the bacteria that cause underarm odor. I checked with the pediatrician first to make sure it would be safe. Sure enough, Listerine works. She applies it after showering, lets it dry, and then applies an antiperspirant. She can go just about the entire day with barely any odor.

A Thanks for sharing this unique solution to a common problem. Listerine contains thymol, eucalyptol, menthol, and methyl salicylate. These oils have antifungal and antibacterial properties. Although Listerine is not approved for this use, we're glad to learn it works.

MILK OF MAGNESIA

Q I want to share a remedy I learned when traveling in Brazil. Just apply milk of magnesia to your armpits. It is the best underarm deodorant!

A What an unusual idea. Milk of magnesia contains magnesium hydroxide, which is both an antacid and a laxative. We have never heard of applying it to underarms, though. Perhaps it reduces the acidity of the skin to make odor-forming bacteria less welcome.

RUBBING ALCOHOL

Q I am 70. When I was 13, my mother told me to use rubbing alcohol as a deodorant. It works. I have tried over-the-counter products but find Mom's advice works like a charm. All my six children use this cheap remedy, and now their families do, too!

A Thanks for sharing your approach. We have also heard from readers who use white vinegar or Listerine under their arms to fight odor.

Other readers are enthusiastic about applying topical milk of magnesia to take care of underarm odor.

VINEGAR OR VODKA

Q I used to work backstage for the wardrobe department in a theater. Actors sweat, and clothes that are not machine washable are dry-cleaned only once a week. Clothes get sweaty and smelly. The solution I was taught is to spray undiluted white vinegar or vodka (the cheaper and higher proof the better) on the armpits and other sweaty areas of the clothing. Once dry, the clothes didn't smell. This worked for the 15 years that I did it.

A Thanks for this fascinating tip. During summer clothes get sweaty quickly. This seems like an affordable solution to a common problem. Before spraying large parts of the garment, though, it might be a good idea to test the vinegar or vodka in an unobtrusive place to make sure it won't stain the clothing.

ZINC OXIDE

Q Years ago I was suffering from sensitivity to all underarm deodorants on the market. I found an alternative product at the health food store and bought it, though it was expensive ($12). The directions said to apply a small amount of this white paste to each clean, dry armpit only once a week. I tried it, and it worked. The ingredients were zinc oxide, rosewater, and some kind of powder. The tin was so small that I used it up in no time. Then I bought a tube of cheap zinc oxide ointment (75 cents) and used it instead. I've been using it ever since. It's not an antiperspirant, but it is a marvelous deodorant. Also, it's safe: Diaper rash cream is made of zinc oxide.

A Thanks for telling us about your experience. According to our cosmetic consultant, zinc oxide has antimicrobial properties. We agree that it is a safe and inexpensive approach to underarm odor.

Brain Function

There's no question about it: Loss of mental ability looms large on the list of health worries, right up there with heart disease and cancer. As the baby boomers age, all three of these chronic conditions are becoming increasingly prevalent. Although there are effective medical treatments for heart disease, and even for certain cancers, those for dementia leave a lot to be desired. Doctors have very little to offer for preventing ordinary cognitive decline, but exercise, social interaction, and diet can be very helpful. What should you eat to keep your brain sharp?

BLUEBERRIES

Q I've heard that blueberries have a beneficial effect on the brain. Can you tell me more about this? Is the research recent and credible?

A James Joseph at Tufts University is a leading neuroscientist and expert on the effects of berries on brain function. His studies on both aging rodents and humans have demonstrated cognitive benefits from blueberries. We think his research is highly credible.[1] Dr. Joseph recommends frozen berries as an economical way to get the antioxidant potential of this fruit.

FISH OIL

Q I keep reading that eating fish is good for your heart, your brain, and just about everything else. I am not fond of fish, so I have started taking fish oil capsules. A friend says fish oil is contaminated with mercury. I hate to think I'm slowly poisoning myself by trying to improve my health.

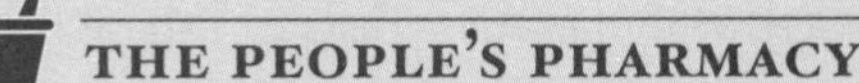

Favorite Food #2: Blueberries

We love blueberries. Actually, we love all berries, whether blue, black, red, or any color in between. Ounce for ounce, blueberries pack a bigger antioxidant punch than almost any other fruit or vegetable. Just 3.5 ounces of blueberries has 2,400 oxygen radical absorbance capacity (ORAC) units. That's about twice as many as spinach or broccoli.

One big helping of blueberries has the same antioxidant power as nearly five servings of certain other fruits and vegetables. And you don't have to spend a fortune on fresh blueberries out of season: Frozen berries work just as well and they are a tasty and economical alternative. Plus, it may be a lot easier to get your kids to eat their blueberries than to convince them to swallow another mouthful of cooked spinach.

Studies indicate that blueberries help brain cells withstand stress.[1] Studies on rats show great results, but even studies on human beings with mild, aging-related cognitive impairment show that people given blueberries or Concord grape juice perform better on tests of verbal memory than people who are given placebos.[2]

Strawberries and walnuts offer some of the same benefits. Other studies have found that the antioxidant compounds in blueberries provide still more benefits. They also can help lower blood pressure[3] and protect against stroke[4] in rats.

Promising work at Ohio State University suggests that blueberry extract can block the development of blood vessels that feed tumors and can inhibit inflammation that encourages their growth.[5] More recently, a small study on mice pointed to a possible protective effect against breast cancer.[6] Researchers gave mice blueberry extract or a placebo every day for a week before injecting the mice with hard-to-treat breast cancer cells. The dose of the extract would be comparable to eating five ounces of blueberries a day. After six weeks, the mice getting the blueberry extract had tumors that were 70 percent smaller and less likely to migrate than the mice getting the placebo.

A — The FDA cautions that large fish (shark, swordfish, king mackerel, and tilefish) may be contaminated with mercury. Tests of fish oil capsules, however, have not revealed mercury. For more details on one such analysis, check out *www.consumerlab.com.*

Bug Bites and Stings

When you are dealing with a sting or a bite, common sense is essential. A person who's had an allergic reaction to a sting may need immediate first-aid treatment with epinephrine (EpiPen) while being rushed to the emergency department. Reactions to a tick bite (generally fever and a rash days or weeks later) also call for prompt medical attention. Lyme disease and Rocky Mountain spotted fever are serious diseases spread by ticks. Usually, though, the pain or the itch is the main issue, and people have found many ways to cope.

BAKING SODA

Q — A wasp stung me today on the inside of my thumb. I called NHS Direct for advice. Then I logged on to your website and found a remedy suggesting bicarbonated soda and vinegar for stings. It worked! Ten minutes later, the pain was nearly gone.

A — American readers may not know that Britain's National Health Service (NHS) provides advice by telephone, digital TV, and the Web. We are delighted that a paste of baking soda and vinegar worked for you. Many other readers have found this home remedy helpful.

CASTOR OIL

Q — I got bitten by fire ants yesterday, and the bites swelled up and itched like crazy. I applied castor oil right away, and the itching stopped.

The bites are just a little bit swollen today, and they don't itch at all. Castor oil sure works on ant bites!

Castor oil is an old-fashioned remedy that people once used as a laxative. We've heard of using it on warts, bruises, and sore joints, but this is the first we've heard of using it on ant bites. Other remedies that readers have used successfully on fire ant bites include baking soda and vinegar, toothpaste, and Vicks VapoRub.

HOT WATER

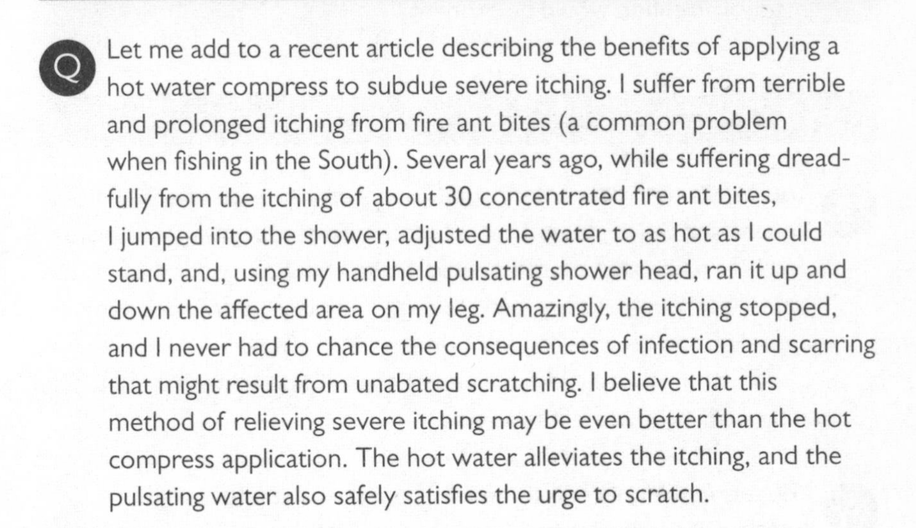

Let me add to a recent article describing the benefits of applying a hot water compress to subdue severe itching. I suffer from terrible and prolonged itching from fire ant bites (a common problem when fishing in the South). Several years ago, while suffering dreadfully from the itching of about 30 concentrated fire ant bites, I jumped into the shower, adjusted the water to as hot as I could stand, and, using my handheld pulsating shower head, ran it up and down the affected area on my leg. Amazingly, the itching stopped, and I never had to chance the consequences of infection and scarring that might result from unabated scratching. I believe that this method of relieving severe itching may be even better than the hot compress application. The hot water alleviates the itching, and the pulsating water also safely satisfies the urge to scratch.

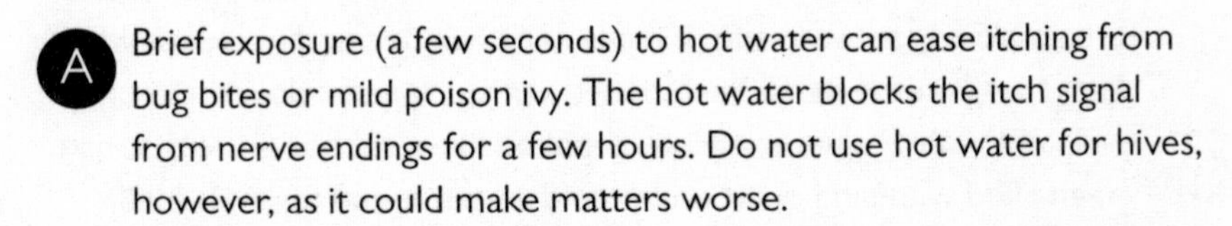

Brief exposure (a few seconds) to hot water can ease itching from bug bites or mild poison ivy. The hot water blocks the itch signal from nerve endings for a few hours. Do not use hot water for hives, however, as it could make matters worse.

MEAT TENDERIZER

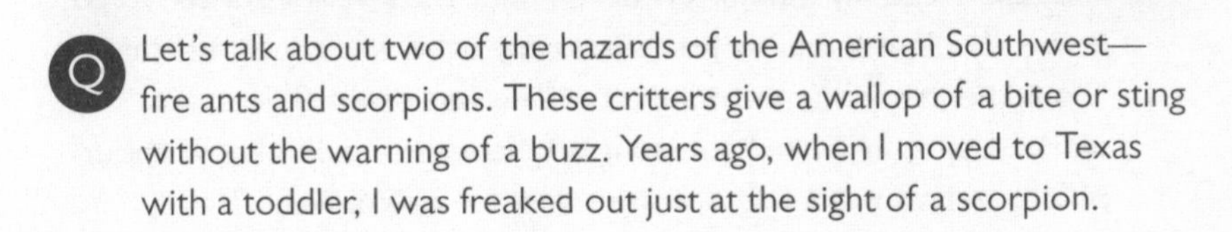

Let's talk about two of the hazards of the American Southwest—fire ants and scorpions. These critters give a wallop of a bite or sting without the warning of a buzz. Years ago, when I moved to Texas with a toddler, I was freaked out just at the sight of a scorpion.

Poison control in San Marcos, Texas, told me to keep meat tenderizer—the kind with papaya extract in it—on hand for bee, scorpion, and fire ant stings. Add a little water to a scant handful of tenderizer and put the paste on the sting immediately. It works! The poison control people also said we should keep tetanus shots up to date because stinging things, especially fire ants, can spread infection.

A The papaya extract (the enzyme papain) in meat tenderizer breaks down proteins—not only in meat, but also in stings. It is a time-honored treatment for bee and wasp stings. Scorpion stings might be too serious for home remedies. If a scorpion stings your child, medical attention would be prudent.

RAW ONION

Q You've written about baking soda and vinegar to ease wasp stings. I once used this remedy. Then a friend suggested applying raw yellow onion to a bite as soon as possible. I take about a teaspoon of grated onion, put it on the sting, and bandage it in place. It provides immediate relief, and the swelling soon disappears. I always take a raw onion as part of my first-aid kit on camping trips. If I don't need it for a sting, I can always use it in a stew.

A You aren't the only one to benefit from raw onion for stings. Another reader posted this story to our website: "I'm a pianist, and I react poorly to wasp stings. I get really worried and freak out when I get stung on the hand. Yesterday I was stung just below the thumbnail, and within minutes my hand looked like a rubber glove that had been filled with air. I put it under cold water, found my Apis Mell (homeopathic remedy for bites and stings), and also took ibuprofen. Then I looked online for help. At your site (*www.peoplespharmacy.com*) I saw people had success with onions, so I cut a slice of onion and taped it to my thumb. Within an hour the swelling started to go down. By dinnertime, six hours later, my hand was almost normal. I could bend my thumb, and the swelling was down. The onion works. Last year when I got stung on the wrist, I didn't know about onion, and my hand was almost useless for over a week."

TAPE

Q I live near a national park and walk my dogs in the woods. There are ticks everywhere. I stop and pull ticks off myself every few minutes, but I hate to just throw them back in the bushes where they will wait for me the next time I go for a walk. Is there an easy way to kill or dispose of them? My dogs have been vaccinated against Lyme disease, but I understand there's no vaccine for me. I surely don't want this disease!

A Put on insect repellent containing DEET before you leave home. Spray your shoes and socks, and tuck trouser legs into your socks. Carry a roll of Scotch Tape in your pocket. Whenever you spot a tick, use the tape to trap it. Once it is sealed in tape, it can't escape.

TOOTHPASTE

Q Your column mentioned toothpaste as a cure for bites from fire ants. Several years ago I was stung many times by yellow jackets. Twelve hours later I was still hurting as though it had just happened.
My daughter had seen a TV show that mentioned toothpaste for stings. I tried it and was surprised to find that it worked immediately. Since then I have used it on various bug bites and stings, including jellyfish stings, with great results. The program stated that it must be mint toothpaste. Perhaps it is the mint that does the trick.

A A number of readers have told us that putting toothpaste on fire ant bites can be very soothing. We have never heard that toothpaste would be helpful against other stings as well. No one seems to know whether it is the mint, the fluoride, or some other ingredient that eases the pain. However, anyone who is allergic to yellow jackets, bees, or wasps should get emergency treatment immediately and not resort to home remedies for stings.

Q Shortly after I read about using toothpaste for fire ant bites, I was bitten by a fire ant. I started spreading toothpaste over my swollen ankle twice a day. It has been three days, and the inflammation has mostly gone away. I only have a small red area around the center of the bite. Previously I would have been at the doctor's office and receiving antibiotics by now. Additionally, the toothpaste helped with the itch.

A Readers have applied many remedies to fire ant bites. In addition to toothpaste, they report success with tobacco juice, Vicks VapoRub, Listerine, apple cider vinegar, and meat tenderizer mixed into a paste with water.

VINEGAR AND BAKING SODA

Q I used a combination of apple cider vinegar and baking soda on a wasp sting and felt instant relief. The foaming action should amuse children who have been frightened by the sting.

A When you combine vinegar and baking soda, the impressive foam is caused by the release of carbon dioxide. We don't know why this combination seems to ease the pain of wasp or bee stings, but it is popular. Other treatments include fresh onion juice or a paste made of meat tenderizer and water.

Burns

For burns, cool the skin immediately with cold water by holding the burn under the tap. Some people plunge a burned hand into a pitcher of ice water or iced tea. This also helps ease the pain of a mild burn. After the cold water treatment, assess whether the burn is serious and requires a trip to the emergency room or whether the problem is less severe and a home remedy is appropriate.

ALOE VERA

Q Some time ago a gardener friend gave me a large potted Aloe vera and said, "If you get a burn, break off a leaf and rub the sap that oozes out onto the burn." I have had a couple of minor burns since, followed the instruction, and experienced relief. Presumably this is not appropriate for severe burns.

A The use of Aloe vera dates to the ancient Egyptians. They used it to treat skin infections and to make laxatives. Today the value of Aloe vera gel for burns is somewhat controversial, although people have used it for this purpose for centuries. Aloe gel contains polysaccharides—compounds that have a soothing effect on mucous membranes—and enzymes that coat irritated skin and ease pain. It also may be antibacterial. In a study of 30 people with second-degree burns, aloe cream containing 0.5 percent of the gel in powdered form helped to heal the burns faster than sulfadiazine, a commonly used antibacterial cream. You are right that serious burns require immediate medical attention.

RAW ONION JUICE

Q While I was working at a restaurant, one of our chefs was burned badly by a fryer. I happened to be in the kitchen when it happened, and the manager screamed, "Get me a fresh onion out of the walk-in refrigerator!" I didn't ask questions. I just got it. He asked me to cut it in half, which I did. He squeezed the fresh onion juice on the chef's burn. Amazingly, it not only calmed the awful pain, but also prevented the burned skin from blistering! The manager later explained to me that it must be a freshly cut onion. I learned the truth of that later when, in another restaurant, I got burned and reached for onions cut up that morning. That didn't work, so then I had to cut a fresh onion. Seconds count when it comes to a burn. But onion juice always works! Something about the chemistry of the juice works wonders.

A Thank you for sharing your experience. First aid for a burn is soaking it in cold water immediately. After that, if the skin is intact, a home remedy such as onion juice might be worth a try. Obviously, a severe burn requires medical attention.

SOY SAUCE

Q I listened to your public radio show and heard a man call in recommending soy sauce for burns. "How weird is that?" I thought. But then, as I took a loaf of bread out of the oven, the inner edge of my thumb and the fleshy pad underneath hit the metal rim of the pan. I expected a painful burn. Since I had nothing else at hand, I decided to try the soy sauce remedy. The pain eased up in less than a minute, the soreness did not materialize, and even the redness went away. It may be weird, but it certainly did work!

A We don't know why this home remedy works but have heard from several people that it does, including an army ranger who told us that U.S. Special Forces medics use soy sauce for combat-related burns.

Q When we were at the beach, a friend burned her hand on a hot handle. (I had just pulled the pan out of the broiler.) I grabbed the soy sauce and had her soak her hand in it after she ran the burn under cold water. She reported relief, and the next day she was fine. I was worried it would blister. She smelled like marinade, but that's a small price. We credited you for the save.

A Thanks for sharing your success with soy sauce. We heard about this remedy for burns from an Oregonian listener to our radio show.

Q Soy sauce for burns works! I was changing the air cleaner in my car, and my metal watchband accidentally arced across a battery terminal. I got a severe burn in the shape of my watchband at the point of contact. I remembered the recommendation of soy sauce for burns. I slowly poured it on the burn for about a minute and had no pain then or afterward. I went back and finished my project.

A We always suggest putting cold water on a burn first. Soy sauce thereafter can help ease the pain from a burn, as you discovered. Anyone who is interested in more details about this home remedy and others will find them at *www.peoplespharmacy.com*. Severe burns require immediate medical attention.

VANILLA EXTRACT

Q I recently read your column about using soy sauce on burns. Vanilla does the same thing. I have used vanilla for many years in my kitchen to soothe burns.

A We always recommend that burns be treated with cold water immediately. Thanks for sharing your vanilla remedy.

Bursitis

Bursitis is an inflammation of the bursa, a sac of fluid inside a joint. The result is pain, and the usual prescription is a nonsteroidal anti-inflammatory drug (NSAID) such as diclofenac or ibuprofen (also widely available without a prescription). While these NSAIDs can certainly ease the pain, not everyone tolerates them well. They can irritate the digestive tract and even cause bleeding ulcers. Bursitis can take a long time to heal. Many people want a home remedy less likely to trigger side effects.

BROMELAIN/PINEAPPLE EXTRACT

Q I have bursitis in my hip. A friend said that you once published a home remedy that helped.

A Another reader recently took us to task for not recommending bromelain for bursitis: "This has worked for many people we know. In just a couple days the pain is gone. Those who continued to take it for a week after that got rid of their bursitis." The enzyme bromelain is extracted from pineapple. Many animal studies have shown that it has anti-inflammatory and analgesic properties. A clinical trial on subjects with painful, arthritic hips found that bromelain (under the brand name Phlogenzym) was just as effective as the NSAID diclofenac (Voltaren).[1]

CURCUMIN

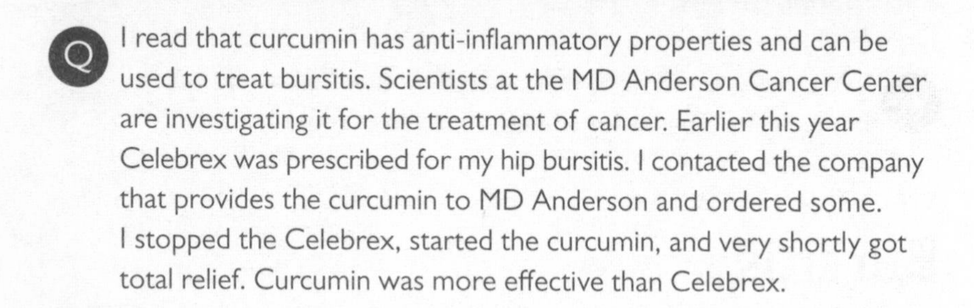

Q I read that curcumin has anti-inflammatory properties and can be used to treat bursitis. Scientists at the MD Anderson Cancer Center are investigating it for the treatment of cancer. Earlier this year Celebrex was prescribed for my hip bursitis. I contacted the company that provides the curcumin to MD Anderson and ordered some. I stopped the Celebrex, started the curcumin, and very shortly got total relief. Curcumin was more effective than Celebrex.

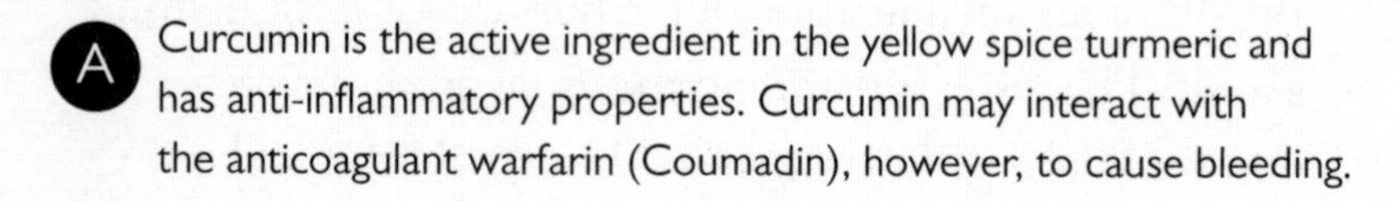

A Curcumin is the active ingredient in the yellow spice turmeric and has anti-inflammatory properties. Curcumin may interact with the anticoagulant warfarin (Coumadin), however, to cause bleeding.

Cancer

There's no simple, straightforward remedy for preventing cancer. Cancer is extremely complicated, and the treatments scientists have developed are also quite complex. But some researchers are starting to recognize the very important role food plays in helping to prevent various types of cancers.

One of those researchers is David Servan-Schreiber, M.D., Ph.D., author of *Anticancer: A New Way of Life* (2008). According to Servan-Schreiber, diet is at the core of a healthy anticancer

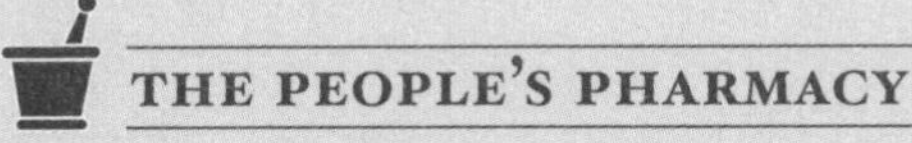

Favorite Food #3: Green Tea

Green tea is typically grown and processed in China, where its history dates back more than 3,000 years. Green tea comes from exactly the same plant—a large shrub called *Camellia sinensis*—as black or oolong tea. The difference is in the way the leaves are treated. With green tea, leaves are steamed immediately before being dried with hot air. This preserves the color and some compounds found in green tea, such as epigallocatechin-3-gallate (EGCG). (To produce black tea, the leaves are "fermented" before being dried.)

Research indicates that regular tea consumption may lower the risk of developing Parkinson's or Alzheimer's disease.[1] Scientists also are exploring possible cardiovascular benefits.[2] Researchers in Texas are interested in using green tea to strengthen bones and to prevent osteoporosis.[3] But most research seems to focus on green tea's potential to reduce cancer risks. An exhaustive Cochrane review—a systematic look at medical research results in a collection of databases—covered 1.6 million participants. The evidence indicates that green tea consumption can lessen the likelihood of developing prostate cancer.[4] Evidence that it can help protect against other cancers is equivocal, although lung, pancreatic, and colorectal cancers may be less common in green tea drinkers. Appropriate dosage appears to be three to five cups daily, which amounts to approximately 250 milligrams of EGCG and related compounds.

The most sensational claim people make about green tea is that it aids in weight loss. While research is inconclusive, substituting green tea for sugar-sweetened beverages can certainly cut calories for dieters. With centuries of use, drinking green tea seems reasonably safe. Green tea does contain caffeine—but less caffeine than coffee or black tea.

One caveat: Patients under bortezomib treatment for multiple myeloma should avoid green tea. Research indicates that EGCG can block the drug's effectiveness.[5] People in cancer chemotherapy should check with an oncologist before adding green tea to their regimen.

THE PEOPLE'S PHARMACY

Favorite Food #4: Pomegranate

If you have never eaten a pomegranate, you have missed something special. No flavor compares to pomegranate's tartness and sweetness, which combine for a culinary delight. For us, this is almost the perfect fruit when it is ripe. The bright red color just adds to its appeal. You can eat the seeds and drink the juice. Best of all, the list of health benefits attributable to pomegranates just keeps growing longer and longer.

Let's start with the antioxidant power of the amazing pomegranate. Research has shown that this fruit can help prevent bad low-density lipoprotein (LDL) cholesterol from oxidizing, which may also reduce the risk of plaque formation and coronary artery disease.[1] Studies show that pomegranates also have an aspirin-like effect on platelets, which circulate in the blood. This antiplatelet effect may reduce the risk of blood clot formation.[2]

Research also demonstrates that pomegranate juice inhibits angiotensin-converting enzyme (ACE) and helps lower blood pressure, thus making it an important part of any heart-healthy diet.[3]

Growing evidence also suggests that pomegranates contain ingredients that have both antibacterial action and antiulcer effects that seem protective for the digestive tract.[4] Research supports the idea that the anti-inflammatory effects of pomegranate may protect against arthritis, diabetes, and perhaps even Alzheimer's disease.[5]

One reader writes that his grandmother even used a pomegranate infusion. She cut the fruit in half, boiled it in water, and then reduced it—to make a tea that effectively treats diarrhea.

Even more exciting are the potential anticancer effects of pomegranate. Researchers have demonstrated that pomegranate "produces anticancer effects in experimental models of lung, prostate and skin cancer. More recently, pomegranate has been found to be anti-carcinogenic in the colon."[6]

Bottom line: We know of relatively few foods that have so much health potential, taste so good, and look so beautiful.

lifestyle. Sugar, white flour, and other refined carbohydrates (all rapidly turned into glucose by the body) can stimulate inflammation and thus, perhaps, cancer cell growth. Based on his research, we have come up with a list of foods that have anticancer potential, listed here from most to least powerful: garlic, leeks, Brussels sprouts, scallions, cabbage, beets, broccoli, cauliflower, onions, kale, spinach, asparagus, green tea, pomegranate, turnips, squash, celery, radishes, eggplant, bok choi, and carrots.

The bottom line is that the more cancer-fighting foods you eat—and the fewer processed foods you eat—the better your chances of remaining healthy and cancer free for longer.

Canker Sores

Doctors have a fancy name for canker sores—aphthous ulcers—but they don't have good treatments. These sore spots on the inside of the cheek or on the tongue usually occur where there has been some kind of small injury from biting the tongue or the cheek, for example, or scratching the inside of the mouth with foods such as pretzels or potato chips. No one seems to know why some people get them and others don't. Canker sores are common, so home remedies abound.

ALOE VERA GEL

Q When I get a canker sore I use the gel from an Aloe vera leaf directly on the sore. I keep a plant handy all the time.

A Aloe gel has a long history of healing cuts and burns, but scientific data proving its effectiveness is slim. However, a small study on licorice-containing patches (marketed under the brand name

Cankermelts) that adhere to canker sores suggests that the sores get smaller and heal faster than untreated aphthous ulcers.

BAKING SODA

Q My husband used to get canker sores. Our dentist recommended using baking soda to brush his teeth. He hasn't had a canker sore since.

A Some people report that avoiding toothpaste containing the foaming agent sodium lauryl sulfate can help prevent canker sores.

Q I have found that toothpaste with sodium lauryl sulfate (SLS) made my mouth ulcers much worse! When I switched to Weleda toothpaste, I saw an immediate improvement. Toothpaste with tartar control seems to be especially irritating.

A Thanks for the testimonial. Some dentists recommend avoiding toothpaste with SLS in order to reduce canker sores.

BANANA

Q I used to get canker sores in my mouth. My mother told me to hold a slice of banana tight against the sore with my tongue, and it works. You have to hold it there until it stings, about four or five minutes. Riper bananas seem to work better. Seldom have I had a sore that lasted longer than a day, and it certainly tastes better than medicine.

A A slice of banana certainly sounds like a pleasant treatment for canker sores. We don't know how it would work.

BUTTERMILK

Q I've read your suggestions for treating canker sores. Some 35 years ago, the school nurse recommended buttermilk to heal them fast!

Why buttermilk might soothe the irritation of canker sores is beyond us, but we cannot think of a safer solution for this annoying and painful problem.

GREEN BEANS

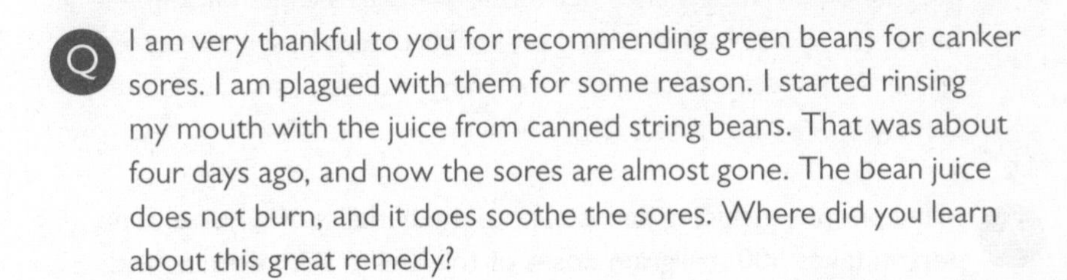

I am very thankful to you for recommending green beans for canker sores. I am plagued with them for some reason. I started rinsing my mouth with the juice from canned string beans. That was about four days ago, and now the sores are almost gone. The bean juice does not burn, and it does soothe the sores. Where did you learn about this great remedy?

A reader wrote in about the value of Gerber's strained green beans for fighting canker sores.

KIWI

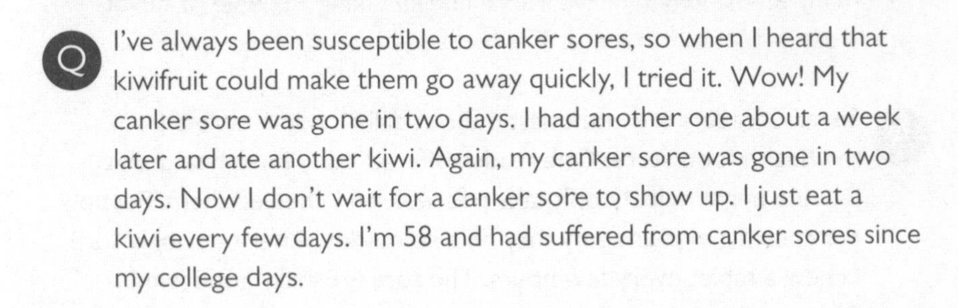

I've always been susceptible to canker sores, so when I heard that kiwifruit could make them go away quickly, I tried it. Wow! My canker sore was gone in two days. I had another one about a week later and ate another kiwi. Again, my canker sore was gone in two days. Now I don't wait for a canker sore to show up. I just eat a kiwi every few days. I'm 58 and had suffered from canker sores since my college days.

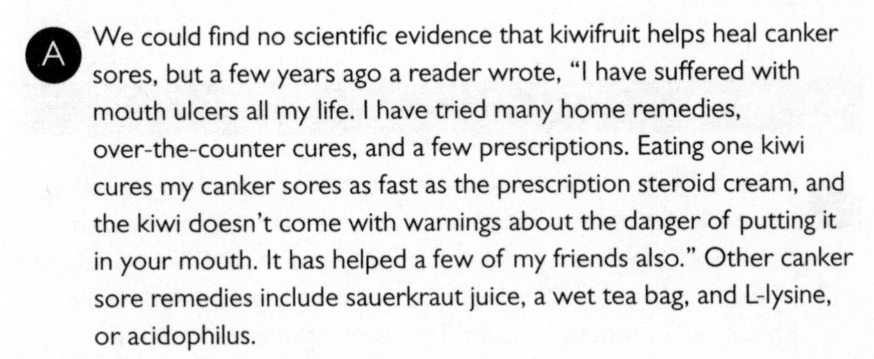

We could find no scientific evidence that kiwifruit helps heal canker sores, but a few years ago a reader wrote, "I have suffered with mouth ulcers all my life. I have tried many home remedies, over-the-counter cures, and a few prescriptions. Eating one kiwi cures my canker sores as fast as the prescription steroid cream, and the kiwi doesn't come with warnings about the danger of putting it in your mouth. It has helped a few of my friends also." Other canker sore remedies include sauerkraut juice, a wet tea bag, and L-lysine, or acidophilus.

L-LYSINE

Q My 37-year-old daughter has battled canker sores her entire life. It's not unusual for her to have more than ten at a time. She's tried a lot of different treatments over the years, with little success. Recently she had a baby and did not have one sore the entire pregnancy. Six weeks after delivering, the sores came back in full force. Her doctors prescribed oral corticosteroid, which cleared up her mouths sores temporarily. They returned as soon as the steroid wore off.

A Many people, including dentists, note the benefits of L-lysine. One person takes 500-milligram doses of this amino acid morning and evening: "Sometimes I'll get a canker sore, but it is gone in two days. I used to get a lot of them, and they would be raw for at least a week."

Q A number of years ago a dentist recommended a possible treatment for canker sores: L-lysine supplements. I have used them successfully many times. When I have a sore I begin taking a couple of tablets a day until the pain eases. The sores seem to disappear with L-lysine.

A Several readers are enthusiastic about daily L-lysine tablets from the health food store. One reader offered this: "I had heard that L-lysine might help. I don't take pills well, so I chewed them. Possibly the chewing explains why they help so quickly. After noticing a sore, I chew a tablet every few hours. The sore is usually gone within 24 hours."

MYLANTA

Q I often suffer from canker sores and have tried everything from two L-lysine tablets daily to nearly every over-the-counter medication. It has always taken the sores quite a long time to heal. Then a retired pharmacist friend of mine suggested swishing around a spoonful of Mylanta in my mouth. Shazam! The sores disappeared quickly.

A Thank you for your recommendation. We have heard of using Mylanta or other liquid antacids for diaper rash, but not for canker sores. Some people report that acidophilus pills also help canker sores heal quickly.

PRELIEF

Q My doctor had me take Prelief because he thought I had interstitial cystitis. It turns out I don't, but I was plagued with canker sores in my mouth every time I ate something acidic, like salsa or barbecue sauce. Since I started using Prelief I have hardly had any. I take it before I eat an acidic food, and it really has helped me. I hope you will pass this information on to other canker sore sufferers.

A Prelief is calcium glycerophosphate. It is sold as a supplement to take the acid out of food, which can be helpful for people who suffer from heartburn or bladder problems. We're glad to hear it worked for your canker sores.

SAUERKRAUT

Q My husband has canker sores in his mouth all the time. The doctors he has seen say there's no cure, so he has to live with it. He has tried many topical remedies, plus some prescription drugs, but he is not sleeping well due to the pain. How can he get relief?

A Here's one reader's remedy: "Sauerkraut juice has worked like a miracle for me! It starts clearing up a canker sore within hours, and the sore usually heals by the next day. Juice from canned sauerkraut doesn't work nearly as well as the fresh stuff (in the refrigerated deli section). It's such a weird remedy that I did some research on sauerkraut to see why it might work. Sauerkraut is fermented (like yogurt or sourdough) and is full of probiotics. I've discovered that if I sip a little sauerkraut juice every couple of days, I don't develop canker sores in the first place."

STYPTIC PENCIL

Q I have suffered from canker sores all my life. Each one resulted in more than a week of pain. I've tried numerous family remedies, but none of them helped. A few years ago a co-worker suggested using a styptic pencil. I put the pencil in a plastic bag and smashed it with a hammer, then put the powder in a bottle with a little water to make a paste. At the first sign of a canker sore, I put some of this paste on it two or three times a day. Usually within a couple of days it goes away completely. This has only one bad side effect: a nasty taste. But I'll live with the taste for two days to avoid seven to ten days of pain.

A Styptic pencils, used to control bleeding from small cuts such as shaving nicks, usually contain alum (aluminum potassium sulfate). We have heard from others that powdered alum can help canker sores heal faster. Alum is sold on the spice shelf of the supermarket. Home canners use it to make pickles crunchy.

Colds

Few things make a person more miserable than a cold. But, while there is no cure yet, there are a few items in your pantry that can help you feel a lot better while you recover. You've probably heard your whole life that chicken soup helps feed a cold. But did you know that there's actually some clinical evidence to back it up? Doctors have been recommending it for centuries—dating to Maimonides in the late 1100s.

Chicken soup fell out of favor with doctors for a time, but in the 1990s, some researchers revisited it. Physician Irwin Ziment hypothesized that the amino acid cysteine, found in chicken soup, might act a bit like acetylcysteine, a drug prescribed to thin mucus in the lungs. Some researchers at Mount Sinai Medical Center in Miami then did tests revealing that cysteine was better at improving mucus flow than hot or cold water. And there's also evidence that chicken soup may slow the white blood cells that cause upper respiratory inflammation.[1] We got our favorite chicken soup recipe from Joe's mother, and we use it whenever anyone in our house shows the first signs of a cold.

Almost as popular as chicken soup is ginger tea. And while you're whipping up hot beverages, another tried and true remedy for adults suffering from cold symptoms is a hot toddy. Okay, so you've heard of chicken soup for colds, as well as toddies and tea. But what about probiotic yogurt? There's some new evidence that probiotics can help to prevent the sniffles.

Of course, it's always good to remember some useful tips for trying to stay healthy in the first place, especially in this era of pandemic flu. First, wash your hands regularly. Second, wash your hands. Third . . . well, you know what we're going to say. But this next recommendation may surprise you—it's been dismissed as an old superstition for years. Some recent data suggest that keeping your feet warm and dry in cold weather may also help fight off viruses and infections. In one study in Wales, subjects whose feet were immersed in cold water were significantly more likely to report cold symptoms in the five days after the experiment than those in the control group.[2]

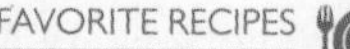

Helen Graedon's Chicken Soup

Large stewing hen and a few additional backs and wings
Onions
Celery
Carrots
Parsnips
Parsley
Bay leaf
Peppercorns
Salt

FOR EXTRA COLD-FIGHTING POWER: *several cloves of garlic (up to a head)*

Cover chicken with water and top with two inches more. Add vegetables and spices. Simmer for about two hours and then strain out the chicken, vegetables, and spices. Remove the meat of the chicken from the bones (careful—it will be hot!), and add it back to the soup with noodles, peas, rice, or other embellishments. Refrigerating the broth overnight makes excess fat removal easy: Just skim it from the top.

Coconut Chicken Soup

Contributed by Sally Fallon Morell

1 quart homemade chicken broth
1 can (14 ounces) whole coconut milk
Juice of one lemon
1 tablespoon grated fresh ginger
1 teaspoon sea salt
Pinch red pepper flakes

TO PREPARE HOMEMADE CHICKEN BROTH: Place leftover carcass of a chicken (bones, skin, and any meat) in large pot, and cover with cold water. Add salt and pepper. Bring water to a boil, and immediately reduce heat to low simmer. Cook uncovered for several hours, skimming foam from surface of broth. Remove bones and strain the broth.

TO PREPARE THE SOUP: Combine all ingredients in a large saucepan. Stir and bring to a simmer. Makes four servings.

Chicken Adobo

Contributed by a listener, from a recipe by his Filipina-American girlfriend

6 pieces chicken
1 cup soy sauce
1 cup white vinegar
6-inch piece fresh ginger root, peeled and sliced thin
1 tablespoon minced garlic
2 or 3 tablespoons brown sugar

Combine ingredients in casserole dish. Bake at 350 degrees for an hour.

THE PEOPLE'S PHARMACY

Favorite Food #5: Chicken Soup

Chicken soup is popular in many culinary traditions and is recognized for its healing powers. Jewish grandmothers are famous for administering this "Jewish penicillin," but traditional cuisines from Mexico to China also include chicken soup as a classic dish. There are variations on the vegetables and on the seasonings, but chicken simmered in water seems a constant.

The medieval physician Moses Maimonides recommended chicken soup for colds. Even today, doctors don't have medicines to prescribe for colds. They might do well to prescribe a piping-hot bowl of chicken soup. It is far less likely to cause side effects than many of the drugstore cold remedies, and it might be more effective.

A couple decades ago, physicians at Mount Sinai Medical Center in Miami pitted chicken soup against plain water, hot and cold. They found that chicken soup greatly improved the flow of mucus through nasal passages. And even though there are prescription drugs to treat influenza, a careful evidence-based review a few years ago concluded that chicken soup can probably help alleviate flu symptoms.[1]

In one study, ingesting chicken soup reduced the inflammation associated with immune system cells called neutrophils.[2] This inflammation contributes to many of the unpleasant symptoms we suffer from colds and flu. Chinese grandmothers add astragalus root to their chicken soup, which may increase the soup's ability to boost the immune system. Some people add garlic.

Pulmonologist Irwin Ziment is a chicken soup enthusiast who advocates adding hot peppers to chicken soup for symptomatic relief of both sinusitis and bronchitis.[3] We also have heard from our readers—and have personally experienced—that chicken soup with thyme can help ease a cough. There is also some evidence that thyme along with other spices that are used to flavor chicken soup, such as garlic, bay leaf, rosemary, and black pepper, may have antimicrobial activity.[4]

GINGER

Q I've heard that ginger is great for both colds and flu. Is there any truth to that?

A Ginger tea is popular the world over for colds. There aren't any hard data on ginger's cold-fighting efficacy, but some animal research suggests that an ingredient in ginger may help fight coughs. And there's certainly no shortage of anecdotal evidence about its ability to ease cold symptoms. One of our favorite recipes came to us from a listener in the foothills of West Virginia; she learned it from her grandmother in India.

Q I love ginger tea for colds and the flu. I add a clove of garlic and a jalapeño pepper. I drink the tea while having a bath, wrap myself up in a robe, and always fall deeply asleep. If I do it at the first sign of symptoms, I don't get a cold (and, besides that, it tastes much better than you'd think!).

A Sounds like a wonderful solution.

Q A few years ago, I had a terrible case of what my doctor called adult whooping cough; literally, I coughed for months. Nothing seemed to help. (Even the Rx cough syrup concoction did not work for more than a few days!) A dear aunt in Hawaii heard my plight and she sent me a ginger tea mix. She said that healers and other Hawaiian islanders use it for colds, coughs, and other respiratory problems. When I drank the tea mixture, it burned all the way down, but it sure worked!

A We've heard from many readers about the wonders of ginger for coughs and colds. Thanks for sharing your experience.

Ginger Tea 1

Contributed by a listener, who learned it from a practitioner of ayurvedic medicine

1 inch fresh ginger root
2 tablespoons lemon juice
Pinch cayenne pepper
2 teaspoons honey or maple syrup

Grate ginger root into a mug. Add ten ounces of boiling water and allow ginger to steep for five minutes. Strain grated ginger out, and add lemon juice and cayenne pepper to liquid. Sweeten to taste with honey or maple syrup.

Ginger Tea 2

Contributed by a listener from West Virginia, who learned it from her grandmother in India

Fresh ginger root
8 ounces boiling water

Peel a fresh ginger root and grate into a mug. Pour water over it. Steep five minutes, strain, and sweeten to taste.

Hot Toddy

1 spoonful sugar
1 cup hot water
1 tablespoon lemon juice
1 shot rum

Put sugar in the bottom of a mug or heat-safe glass. Add hot water, lemon juice, and rum. Stay home, sip, and relax.

PROBIOTIC YOGURT

Q Lately I've heard a lot about probiotic yogurt being good for colds. I'd like to know more about probiotic yogurt.

A A new study out of China provides some very promising evidence that probiotics may help prevent colds and the flu, at least for children aged three to five. In the study, those who drank milk with *Lactobacillus acidophilus* were half as likely to get a cold or fever as those who drank plain milk. Kids drinking milk containing *Lactobacillus* plus the bacterium *Bifidobacterium animalis* were 72 percent (!) less likely to develop a fever.[3]

Cold Sores

Cold sores can be painful and unsightly. They are caused by a virus (usually herpes simplex virus 1). Recurrent cold sores can be treated effectively with prescription antiviral medicines such as acyclovir (Zovirax), famciclovir (Famvir), or valacyclovir (Valtrex). Many people prefer a less expensive remedy, however.

BUTTERMILK

Q A few weeks ago a reader wrote to you about cold sores. This person derided the use of "silly" remedies like lysine. I know from experience that lysine works, but plain old buttermilk works just as well and even more quickly.

A Physicians, pharmacists, and housewives have all written to tell us that taking the amino acid lysine and drinking buttermilk both work to prevent cold sores (fever blisters) or canker sores (aphthous ulcers). There are no placebo-controlled trials, but this seems like an inexpensive and low-risk approach.

L-LYSINE

Q You had an article about drugs that heal cold sores faster. Why not prevent them? I take L-lysine before meals every day to prevent cold sores from developing into blisters. Prevention is better than cure.

A Several studies suggest that the dietary supplement L-lysine may reduce cold sore outbreaks.[1] The optimal dose is unknown, but side effects seem rare. Some people contend that limiting the amino acid arginine in the diet improves the effectiveness of L-lysine. Foods high in arginine include chocolate, nuts, and seeds.

Q I get cold sores and was told to take L-lysine daily to help prevent them. I take one pill per day and have not had a cold sore in five months! Is L-lysine dangerous?

A L-lysine is an amino acid, a building block for protein. Several studies have shown that it prevents cold sores. It appears quite safe, except for people with kidney or liver disease.

Colitis and Crohn's Disease

Colitis, also known as inflammatory bowel disease, is hard to treat and difficult to endure. Anyone with inflammation of the colon should seek a specialist's care. Related to colitis, Crohn's disease might be considered a type of colitis except that any part of the gastrointestinal system can be involved, not just the large bowel. In both inflammatory conditions, the immune system runs amok and attacks the digestive tract. Scientists still don't understand the causes nor do they have a cure. Foods or home remedies are not substitutes for medical treatment. Nevertheless, you will be surprised at some of the success stories you read below.

THE PEOPLE'S PHARMACY

Favorite Food #6: Pineapple

Think of Hawaii, and chances are you will think of this delicious tropical fruit, known by the botanical name *Ananas comosus*. Nutritionally, the pineapple is rich in vitamin C and manganese, but the medicinal effects probably come primarily from its bromelain content.

Bromelain is the compound responsible for pineapple's ability to break down protein. (This is why fresh or frozen pineapple added to a gelatin mold will turn it to soup. Cooks use canned pineapple in Jell-O for this reason.) Bromelain can do far more than dissolve gelatin, though. Early on, it was reputed to be a beneficial digestive enzyme, perhaps similar to the papain derived from papayas.

More recent research in mice has demonstrated that bromelain can calm inflammatory bowel disease. It does this by affecting how white blood cells respond to inflammatory triggers.[1]

Many readers have reported that bromelain can also be helpful for inflamed joints. This may be due to its ability to inhibit COX-2, the same enzyme that is blocked by the prescription arthritis drug Celebrex.[2] However, when put to the test in people who experience moderate to severe osteoarthritis in their knees, bromelain came up short.[3]

Nevertheless, according to reported studies, scientists have found some intriguing anticancer properties of bromelain.[4] It's pretty unlikely that simply eating pineapple will be able to ward off cancer, but given this basic research, it makes sense to include pineapple among the fresh fruits eaten as part of an overall healthful diet.

THE PEOPLE'S PHARMACY

Favorite Food #7: Coconut

Coconuts defy logic. How did somebody see that hard, hairy specimen and imagine something delicious inside? It may be that people living in the tropics where coconuts grow (and where travelers are often visited by "Montezuma's revenge") quickly made the discovery that coconuts are excellent for soothing digestive upset.

In fact, one of our readers told us, "My grandmother was the medicine woman in our village [in Nicaragua]. She used to give me coconut milk for diarrhea or parasites, the younger the coconut the better." Over the years, we've heard of coconut remedies for everything from traveler's diarrhea to irritable bowel syndrome to Crohn's disease. (Always keep in mind, however, that eating too much coconut can lead to constipation.)

Coconut comes in many delicious incarnations. (Coconut cream pie, anyone?) Gluten- and dairy-free coconut macaroons are made by several companies, including the ever-popular Archway and Jennies brands. Two cookies per day seem to decrease diarrhea dramatically. If you have trouble finding this product in your local grocery or online, you can always bake your own batch. Other coconut treats include Mounds bars and coconut ice cream or sorbet.

But people watching their sugar consumption or their weight—two cookies per day can add a lot of extra calories, after all—don't have to take their medicine in the form of a dessert. Sprinkle shredded coconut into yogurt or oatmeal, or sip coconut milk.

Coconut water—Vita Coco, O.N.E., and ZICO brands, for example—is also becoming more widely available in the United States. We haven't yet heard from anyone who has tried coconut water to calm a wrathful GI tract. But coconut water is lower in calories than any other coconut-related option and may be worth a try if concerns about your gut involve more than what's happening inside it.

COCONUT

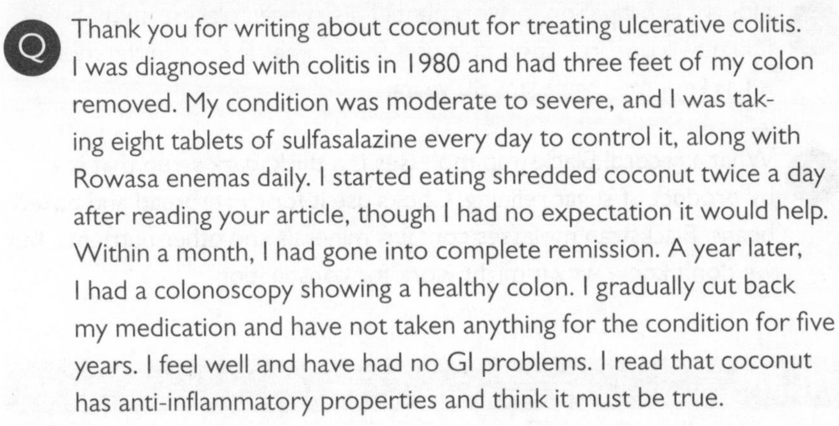

Q Thank you for writing about coconut for treating ulcerative colitis. I was diagnosed with colitis in 1980 and had three feet of my colon removed. My condition was moderate to severe, and I was taking eight tablets of sulfasalazine every day to control it, along with Rowasa enemas daily. I started eating shredded coconut twice a day after reading your article, though I had no expectation it would help. Within a month, I had gone into complete remission. A year later, I had a colonoscopy showing a healthy colon. I gradually cut back my medication and have not taken anything for the condition for five years. I feel well and have had no GI problems. I read that coconut has anti-inflammatory properties and think it must be true.

A We have heard from others who suffered from severe diarrhea associated with inflammatory bowel disease that coconut macaroon cookies or shredded coconut alone can be helpful. We doubt that many people with ulcerative colitis would respond as dramatically as you have, but we certainly are delighted to learn that this remedy was beneficial for such a serious condition.

Constipation

Americans love laxatives, but they're not alone. The medical traditions of many cultures focus on regularity as essential to good health. We can't prove that claim one way or the other, but constipation can be uncomfortable. Of course, a regular pattern may be a bowel movement twice a day for one person and three times a week for another. We are more enthusiastic about natural ways to deal with the problem than constant use of laxatives. Abuse of laxatives—especially the "natural" laxatives, like senna or cascara sagrada—can compound the problem. Fortunately, many home remedies can help ease constipation.

BLACKSTRAP MOLASSES

Q I fought constipation for years until I discovered blackstrap molasses. I take two or three spoonfuls of it three times a week in the morning. It has kept me regular for 40 years.

A What a record! Blackstrap molasses is a thick, dark syrup that is a by-product of sugar refining. Cooks use it for gingerbread and baked beans. Blackstrap molasses contains minerals and other nutrients, but we don't know why it might work for constipation.

FIBER

Q I was troubled with irregularity most of my life until I discovered Fiber One cereal. I take it with me everywhere—on cruises, to Europe, even to the hospital. Fiber One bars come in handy when milk is unavailable. It's the only remedy that's ever worked for me. I can't stand liquid Metamucil, so I tried the capsules with no success. Benefiber also did nothing for me, so I'm a Fiber One fan forever.

A Thanks for sharing your solution. Others may also benefit from psyllium (Metamucil and other brands), wheat dextrin (Benefiber), or cellulose (Citrucel).

Q Over the years I've had issues with constipation, and I recently discovered a wonderful aid. In an attempt to increase the fresh vegetables I eat each day, I have been roasting various veggies—eggplant, squash, zucchini, carrots, and thick slices of onions. I toss the cut-up veggies with garlic and pepper seasoning and olive oil and roast them in a 425-degree oven for about 30 minutes. This solved my problem.

A Thanks for the suggestion on a tasty way to get more fiber into the diet. Eating lots of vegetables is likely to be good for anyone, whether or not it banishes constipation. Fluid and fiber are the first line against this common problem.

Q I suffer from constipation when I travel. The minute I get on a plane, things stop. I go to Europe two or three times a year, and the misery is beyond words. Since I suffer from this problem at no other time, I can't try remedies at home, and I have shied away from laxatives. I am going to London soon, and believe me, it is on my mind already!

A No gastroenterologist has explained to us why traveling creates constipation in so many people. We suggest you take along Metamucil cookies. They contain soluble fiber and are a lot more convenient than powdered psyllium seed. Sugarless gum can also help. Flying dehydrates the body, so drink lots of water on the flight.

FLAXSEED OIL

Q I have found a remedy that has worked well for inflammatory bowel disease (IBS), which caused constipation that put me in the hospital about twice a year in excruciating pain. They would load me up on pain meds, run tests, and eventually release me, only to have it happen again and again without any warning. Then I started taking flaxseed oil capsules at least once a day. This has virtually cured my constipation, and I have had no IBS problems since.

A Ground flaxseeds, such as those found in Uncle Sam Original bran cereal, are also quite helpful in preventing constipation. Another reader reports that two tablespoons of flaxseed simmered for 15 minutes in three quarts of water produces an anticonstipation tonic. She strains and refrigerates the liquid and takes two ounces in juice every morning for regularity.

HONEY AND VINEGAR

Q I have a concoction for constipation that may help others: Mix two teaspoons honey, one teaspoon vinegar, and enough hot water to make four ounces. This is the first thing I drink every morning. It helps my stiff joints, too.

 "Hot lemonade" made with lemon juice, honey, and hot water is a time-honored morning beverage said to encourage regularity. Your drink sounds like a variation on that theme.

MAGNESIUM

Q I was constipated for five years. I was using Citrucel and extra bran on my cereal. I drank lots of water but still had hard, rabbit-like stools. A friend suggested magnesium. It has worked wonders. I take 500 milligrams before bed and have a good response, usually before noon the next day. My internist and cardiologist assure me it is safe.

A Magnesium has long been used to counter constipation. (Milk of magnesia is a well-known laxative.) But too much can cause diarrhea. Most people tolerate 300 milligrams with no problems, but those with kidney problems must avoid extra magnesium.

POWER PUDDING

Q You recently shared a nurse's power pudding recipe of bran, applesauce, and prune juice for constipation. How much of this mixture do you recommend as a daily dose?

A The recipe calls for one cup of wheat (not oat) bran, one cup of applesauce, and three-quarters cup of prune juice. Take one or two tablespoons daily, washed down with lots of water. The glop, which is quite stiff, should be kept refrigerated. It is an excellent remedy for hard-to-treat constipation, but it is critical to get plenty of fluid to avoid intestinal problems.

SUGAR-FREE SWEETS

Q I have read about constipation treatments in your column. My solution—ice cream with sorbitol (the nonsugar sweetener). It has worked for me for years.

A Nonsugar sweeteners like sorbitol are not absorbed from the digestive tract and have a laxative effect. Whether found in sugarless gum, candy, or ice cream, they can all help relieve constipation. Too much, though, may cause diarrhea.

Q I read with interest and sympathy a letter about problems with constipation. I wanted to share something that has helped me. After hearing complaints that sugar-free jellybeans cause diarrhea, I tried them to see if they would help my constipation. I found that if I eat 30 sugar-free jellybeans with a glass of water half an hour before bedtime, I stay regular. I hope this helps others with the problem.

A Thank you for the tip. Many people find that sweeteners in sugar-free candy can cause diarrhea. How clever of you to turn that side effect to your advantage! Each person will have to experiment to find the right "dose."

Coughs

Coughs can be infuriating, and occasionally serious. A cough due to a serious bacterial infection or to chronic acid reflux should never be ignored. But many coughs start as a symptom of a cold or other viral infection. There's not much medication that will be helpful in such cases. In fact, research has shown that the usual cough medicines are no more effective than placebos for children with colds. For children older than one, honey may work better than a drugstore medicine, and with less risk of side effects. (*Note:* Do NOT give honey to babies one year old or younger! They may not be able to handle botulinum sometimes found in honey.) Readers have come up with many home remedies to ease coughs.

CONCORD GRAPE JUICE

Q My wife used to get sore throats every winter. They'd hang on for weeks and develop into a loud, hacking cough. Until she recovered, neither of us would get much sleep. Then I remembered a remedy my sister recommended: drinking grape juice regularly. Both my wife and I started drinking a glass of Concord grape juice every day, fall through spring, and the problem vanished. Since then, we've almost never had a sore throat or bad cough. We drink half a glass of grape juice and add half a glass of water. We make the juice from frozen concentrate. Do you know why this works?

A Purple grape juice has a surprising number of potential health benefits. Research has shown that it can reduce bad cholesterol, lower blood pressure, and help keep blood vessels flexible. Some data suggest that ingredients in grapes boost the immune system. Whether this would help ward off sore throats and coughs we do not know.

THYME

Q I need advice on coughs. My husband has been coughing nonstop, and nothing seems to help him. Can you offer a suggestion?

A It is important for your husband to identify the cause of his cough. If it is related to an infection, his doctor will need to treat it appropriately. Some medicines for high blood pressure (angiotensin-converting enzyme, or ACE, inhibitors such as enalapril, lisinopril, or ramipril) can trigger coughing. If there is no obvious cause, your husband may benefit from thyme tea. Thyme has compounds that can calm a cough. Oils in thyme (thymol and carvacrol) reduce irritation in the respiratory tract. People have used the herb for centuries to loosen mucus and relax muscles that spasm during coughing. Just steep a teaspoon of dried leaves in a cup of hot water for five minutes. Sweeten to taste. Thanks for an interesting modification on an old remedy. Spices like thyme and ginger have a long history against coughs. They also have strong flavors and your solution is a way to make them palatable.

Thyme Cough Syrup

Contributed by Tieraona Low Dog

1 cup near-boiling water
2 tablespoons dried thyme (or 4 tablespoons fresh)
1 teaspoon lemon juice
½ cup organic honey

Pour water over thyme and steep for ten minutes. Strain. Add honey and lemon juice. Refrigerate for up to one week. For children 18 months and older: Take one tablespoon as needed. For those who don't like the flavor of thyme, you can substitute fennel seed. Simmer the seeds gently on low heat for 15 minutes, and then strain.

One reader told us: "I don't like the taste of thyme and ginger in a tea for relieving a cough. But if you add a low-sodium chicken bouillon cube instead of sugar, you have a tasty little broth that calms a cough."

Here's the recipe:

Bouillon cube
1 cup hot water
½ teaspoon thyme
1 teaspoon grated fresh ginger root

Dissolve bouillon cube in water. Add thyme and ginger root. Steep the mixture for four or five minutes, and then pour it through a strainer into a clean mug. Sip and enjoy!

VICKS VAPORUB

Q A few weeks ago my granddaughter spent two nights coughing and kept both her parents awake. I told them of your suggestion to smear Vicks VapoRub on the soles of the feet to stop coughing. She resisted until her dad and older sister agreed to rub it on their feet too. Lo and behold, the coughing child slept all night, and so did the rest of the family.

A Thanks for sharing your story. We don't know why putting Vicks on the soles of the feet is so helpful, but it may have something to do with the anticough activity of menthol. Be sure to put on warm socks over the Vicks to keep the sheets from getting greasy.

ZINC

Q My high blood pressure makes it hard to find safe cold or cough medicines. Black elderberry extract and zinc did the trick.

A Elderberry-flower tea is a traditional remedy for colds and coughs. Many herbalists believe elderberry is more effective than echinacea. Studies of zinc as a cold fighter have produced mixed results. Neither remedy should increase blood pressure, though.

Cuts and Bruises

Minor cuts and bruises seem to be an inevitable part of daily life. While serious injuries may need medical attention, lots of lesser problems respond quite well to home remedies.

ARNICA GEL

Q Yesterday I had a slow-motion tumble off my bike onto my knees on the way home. My knees were bruised, but I was able to pedal home and get arnica gel on them in about ten minutes. I also took some homeopathic arnica pills. My mother swears by this herb, but I had not had occasion to use it much. I applied more gel before bed and again in the morning. Today one knee shows no effects of having been bruised, and the other is not black and blue and barely hurts at all. I am stunned by the effectiveness of this arnica. Have studies been done?

A *Arnica montana* is a flowering plant that grows in Europe. People have traditionally used arnica salves, ointments, gels, and creams for bruises and sprains. Arnica has a long history of use in Europe and is becoming a common addition to first-aid kits in the United States. There is relatively little scientific evidence to support its use, but your report is not the first we've heard. In a recent study, homeopathic arnica pills seemed to have a small but measurable effect on bruising following a face-lift.[1] Homeopathic tablets contain very little arnica. At higher doses, though, arnica should not be taken orally since it can be quite toxic.

BLACK PEPPER

Q Last night I sliced my finger on the inside of a can that I was rinsing out. It was a fairly significant cut. I rinsed it and applied pressure, then went to the cupboard and pulled out a packet of black pepper (which had been stored in the cupboard specifically to treat cuts). I put black pepper on the bleeding cut and then bandaged it, and it seemed fine. After a few minutes, I started to think maybe it wasn't such a good idea to put pepper into an open wound. I decided to rinse out the cut, make sure it was clean, then wrap it again and start over. But when I took the bandage off, the bleeding had stopped and the cut was basically sealed. I rinsed it and put a new bandage on. This morning I looked again, and it hardly looks like there was a cut at all, though it's still quite tender. I'm amazed this worked so well!

A Thanks for your story. Other readers have reported that ground black pepper does seem to help stop bleeding, but a serious cut requires medical attention instead of home remedies.

CASTOR OIL

Q My mother-in-law swears by castor oil for bruises. Whenever one of my kids bumps a knee, she rubs castor oil on it and there is never a bruise. Have you ever heard of using castor oil for bruises?

A Arnica and calendula are herbs that have long been used to relieve muscle aches and to prevent bruises after minor injuries. Castor oil is used internally as a laxative, but many people tell us it works externally for warts. Now we will add bruises to the list.

Q Whenever my children hurt themselves, the first thing I reached for was the castor oil. My mother-in-law's uncle was a boxer and always used castor oil after a boxing match to prevent hematomas and bruising. It works like a charm. It has been passed down as a remedy in my family for many years.

A Thanks for sharing your family remedy. We have heard from others that applying castor oil after a bump can often avert a bruise.

CAYENNE PEPPER

Q A freak razor blade accident sliced the edges of my nose. The bleeding wouldn't stop. Before going to the emergency room, I checked "Stop Bleeding" on the computer. I read that a paste of cayenne should stop bleeding in ten seconds, it said. Mine took 15 seconds.

A We have heard from many people who have used ground black pepper to stop bleeding. Although we also have heard that cayenne pepper works, your story is the first to describe success. We assume you made the paste by mixing it with water. High-tech solutions for minor cuts can be purchased in pharmacies. Look for products such as QR Powder, QuikClot Sport, or BloodSTOP.

SAGE

Q More than 40 years ago I worked in a Chinese restaurant. One day at work I somehow stabbed an ice pick through the end of my thumb, and it bled severely. I couldn't even get it stopped long enough to put on a bandage. Finally, I went into the kitchen to show it to my boss. He took one look, reached for a can of ground sage, and applied it to the wound. I never saw anything stop bleeding as quickly as that! Black pepper isn't the only kitchen remedy for bleeding.

A Thanks for this unusual remedy. We have had firsthand experience using black pepper to stop bleeding from a minor cut. It's helpful to know that ground sage works too. Of course, serious cuts require medical attention.

Dandruff

What is the most annoying thing about dandruff—the flakes all over your shoulders or the itchy scalp? Dandruff is hardly ever a serious medical condition (except the kind caused by psoriasis), but it can be persistent and extremely annoying. It seems to be triggered by an imbalance of yeast living on the scalp. Readers have shared a number of inventive home remedies that may help.

LISTERINE

Q Have you ever heard of using Listerine for dandruff? Someone told me that he heard on the radio how Listerine helps to get rid of the embarrassment of flaky scalp.

A A gentleman called in to our public radio show with an amazing story about Listerine mixed with baby oil. His veterinarian had recommended this combination for relieving itchy spots on his Dobermans and horses. He found that it worked and tried it for his own dandruff. He said it gets rid of dandruff in two to three days. Of course, this is not scientific evidence. But Listerine does contain a number of essential oils (thymol, eucalyptol, menthol, and methyl salicylate) that may have antifungal properties. Since dandruff appears to be caused in part by a fungus (yeast), it stands to reason that a fungus fighter could provide benefit. Other ingredients in dandruff shampoos also counteract fungi. Selenium sulfide, zinc pyrithione, and the antifungal drug ketoconazole (Nizoral) are all effective fungus fighters. The caller did not tell us the precise ratio of Listerine to baby oil, so if you want to try it, you will have to experiment.

Q A few weeks ago you wrote about someone using Listerine and baby oil to treat dandruff. More than 40 years ago, my family was using plain Listerine for this purpose. It was advertised as a dandruff treatment during World War II, and it worked wonders.

A We were fascinated to discover a 1943 ad for Listerine Antiseptic against "infectious dandruff." These days Listerine is advertised only for oral hygiene.

MILK OF MAGNESIA

Q I have been using milk of magnesia on my face for seborrheic dermatitis for the past two months, and my face flakes are gone! I pour it in my hand, massage it on my face (forehead, eyebrows, around the eyes, nose, cheeks, and chin) while showering, and rinse it off at the end of the shower. End of problem. It's a great, cost-effective alternative to expensive Nizoral, and it works better.

A Seborrheic dermatitis is like superdandruff, causing flakes and itching. It appears not only on the scalp, but also on the face. We are glad to hear the milk of magnesia solved the problem.

MIRACLE WHIP

Q For several months now, I have had a problem with a scaly, flaky scalp. I finally went to the doctor, who prescribed medicated shampoo and mometasone topical solution. Nothing worked. Then a friend suggested Miracle Whip. I rubbed it into my scalp and left it in for a couple of hours. I now have a flake-free scalp. Any idea why?

A We checked the ingredients in Miracle Whip. They are water, soybean oil, vinegar, high-fructose corn syrup, sugar, modified food starch, egg yolks, salt, mustard flour, artificial color, potassium sorbate, spice, paprika, natural flavor, and dried garlic. We can't imagine why any of these compounds would clear the flakes from your scalp, but others have praised Miracle Whip as a good hair conditioner.

VINEGAR

Q A few weeks ago someone reported success using Miracle Whip on a scaly, flaky scalp and wanted to know why it worked so well. I think the effective ingredient is vinegar. It's lots cheaper and less messy than Miracle Whip. There are many Web posts on this and I've found from personal experience that vinegar works great.

A Other readers agreed that a dilute vinegar rinse helps reduce dandruff. One suggested mixing four parts warm water to one part apple cider vinegar and using this solution to rinse hair after shampooing. She is 80 and has been using this remedy successfully for 55 years. Other readers use equal amounts of water and vinegar.

Q All my life, I have used dilute vinegar to rinse my hair after shampooing. It works on dandruff and on feet to stop odor. And it's cheap!

A A number of people have confirmed that vinegar (before or after shampooing) is effective against dandruff. The acidity makes it difficult for yeast to thrive. Dandruff and foot odor are both caused in part by yeast that lives on the skin.

Q My wife reads your column and told me about a vinegar rinse to control dry scalp. I have suffered from this problem for years. I have used a huge variety of shampoos, including expensive prescription ones. Sometimes my itchy scalp made it difficult to sleep. The vinegar mixed with an equal amount of water has made a huge difference. Thank you! I have even started rinsing my dog's coat with this solution after a bath. Some areas where his coat was thin have grown back. We spent hundreds on vet bills, and I am pleased to have solved this so inexpensively. I can't thank you enough.

A Many people report that vinegar fights dry skin. Some find that rinsing the hands in a vinegar solution is helpful. The vinegar rinse might also discourage yeast that lives on the scalp and causes dandruff, seborrheic dermatitis, and intense itching. We're glad it helped.

Diabetes

The body runs on sugar. Everything we eat eventually breaks down into glucose, which the hormone insulin carries into our cells. When something goes wrong at any point in this process, the cells don't have enough fuel to run on, and the result is serious illness. In type 1 diabetes, the pancreas stops producing insulin, so the disease is controlled with insulin injections. In type 2 diabetes, which is far more common, the cells lose their ability to respond to insulin. This is known as insulin resistance. Type 2 diabetes can sometimes be controlled through exercise and diet. A number of foods may have a potentially beneficial effect on the glucose-insulin-fuel process. Any food that improves insulin sensitivity might be useful. Diabetes requires collaboration between physician and patient. Diabetics must pair consistent blood sugar monitoring at home with regular measurement of hemoglobin A1c (HbA1c)—an average of blood sugar control over several weeks—at the doctor's office to make sure their blood sugar control is adequate. A surprising number of natural remedies can help people with type 2 diabetes by preventing a rapid rise in blood sugar after eating.

BITTER MELON

Q I have type 2 diabetes and high blood pressure, and I take medicine for both. I have heard that bitter melon can be used to treat type 2 diabetes. What do you know about this plant?

A Bitter melon (*Momordica charantia*) is widely used as both food and medicine in India and China. Studies show that it can lower blood sugar in animals and in human diabetics.[1] Use of any herb or dietary supplement requires careful monitoring and must be coordinated by your physician.

CINNAMON

Q I have high cholesterol and diabetes (controlled through diet). I've been using cinnamon to help keep my blood sugar and cholesterol down. I'd like to continue, but I read in your column that it might be dangerous. Is there a specific brand or type that does not have the damaging ingredient in it? I hate to buy yet another expensive supplement when cinnamon is so readily available in the spice aisle of the grocery store.

A Research shows that cinnamon may help control blood sugar, but German regulatory authorities warn that the kitchen spice sometimes contains high levels of coumarin. This compound can damage the liver or kidneys, if taken at fairly high levels daily. It may also have blood-thinning activity and interact with the anticoagulant warfarin (Coumadin). Cinnulin PF is a water-soluble extract of cinnamon that does not contain coumarin. Capsules, available in health food stores, appear to be safe and may help control blood sugar levels.

THE PEOPLE'S PHARMACY

Favorite Food #8: Cinnamon

Cinnamon comes from the bark of two South Asian trees: *Cinnamomum verum* and *Cinnamomum cassia* (commonly called cassia cinnamon). The cinnamon found on the spice shelf at the grocery store usually comes from *C. cassia*. Active compounds in cinnamon are water-soluble polyphenols, which contribute to the color, taste, flavor, and medicinal actions of many plants. Polyphenols dramatically boost the action of insulin.[1] In a placebo-controlled trial lasting 40 days, researchers in Pakistan gave participants cinnamon capsules in doses of one, three, or six grams. The researchers reported that cinnamon reduced fasting blood sugar—along with triglycerides, LDL cholesterol, and total cholesterol—in people with type 2 diabetes.[2] One study clarified that cassia cinnamon affects blood sugar, while Ceylon cinnamon, from *C. verum*, does not.[3]

Q When my daughter learned that I was diagnosed with type 2 diabetes, she did some research and found out that cinnamon capsules would be helpful. I have used cinnamon for about three years. My family doctor does blood tests and has confirmed that cinnamon keeps my blood sugar under control.

A Research supports your experience. Cinnamon can keep levels of blood glucose from spiking after a meal. We don't recommend using cinnamon from the spice rack, though, since some brands may contain coumarin, an ingredient that can be toxic when ingested in large amounts. Cinnamon capsules are safer.

Q I tried making a cinnamon extract with hot water to help with blood sugar. I ended up with a gooey glob. Please provide exact proportions of spice to water so I don't have to deal with the mess.

A Research shows that one-quarter to one-half teaspoon of ground cinnamon before a meal can reduce the rise in blood sugar after eating. We suggest putting this amount of cinnamon in a paper coffee filter and pouring a cup of hot water over it. One reader has a slightly different technique: "I put about two teaspoons cinnamon in my coffee filter and then put coffee grounds on top so I get the benefits of cinnamon and it cuts any bitterness from the coffee. I turned all my family and friends on to this, and my mother-in-law was able to go off her diabetes medicine that she'd been on for years!" Note: Two teaspoons of cinnamon is enough for a whole pot of coffee. Anyone who uses cinnamon to lower blood sugar should be under medical supervision to have both HbA1c and liver enzymes monitored.

COFFEE

Q My father and uncle both have diabetes. I want to reduce my chance of developing this disease, and I've heard that drinking coffee can help. Is there any evidence for this claim?

A Several epidemiological studies have demonstrated an association between regular coffee consumption and a reduced risk of developing type 2 diabetes.[2] Do not count on coffee alone to protect you, however. Regular exercise and weight control are far more likely to be helpful in preventing type 2 diabetes.

FENUGREEK

Q What is fenugreek? I have been diagnosed as borderline diabetic. My neighbor said this over-the-counter product helps keep blood sugar in check. My doctor said with proper diet and exercise I can beat the diabetes. Do you have any additional information?

A Fenugreek is an herb used in Indian cooking. Research shows that it can help lower blood sugar. Other natural substances used to control blood sugar include cinnamon, bitter melon, oolong tea, and vinegar.

MUSTARD

Q I have been using cinnamon to help control my blood sugar for the last four years. Using one-quarter teaspoon in boiling water to make cinnamon tea lowers my blood sugar readings from about 185 to 135 in one hour. Yellow mustard works even better. I take about one-half teaspoon per meal, depending on the amount of carbohydrates in the food. Both cinnamon and yellow mustard can be overdone and lower blood sugar too much, so you have to be cautious.

A Research supports the idea that cinnamon can lower blood sugar. Until your email, we had not heard that yellow mustard could do much the same thing. Several animal studies show that curcumin, the active ingredient in the yellow spice turmeric, lowers blood glucose. Since turmeric gives mustard its yellow color, perhaps this explains the benefit you have discovered. The vinegar in mustard may also help. However, diabetics should monitor their blood sugar closely and check with a physician before trying such dietary strategies.

NOPAL CACTUS

Q Friends who go to Mexico each year tell me they take nopal capsules to lower cholesterol and sugar in the blood. (They have type 2 diabetes.) Apparently this herb is popular in Mexico. Do you have an opinion on nopales for these conditions?

A We recently heard from a physician that one of his type 2 diabetic patients was able to get control of his blood sugar by drinking a tea made from nopal cactus in combination with his medicine. Nopal is prickly pear (*Opuntia*) and has been studied in Mexico for its ability to lower blood glucose and cholesterol. Research on animals shows it can be effective, but investigators in a human study had to use high doses of nopal capsules to produce results. No one should substitute nopal capsules for diabetes or cholesterol medicine. Diabetics must monitor blood sugar closely if they add any nonstandard remedy to their regimen.

Q My husband has had problems with blood sugar. When he had to go on prednisone I worried because this drug can make blood sugar problems worse. The doctor didn't seem concerned, but as a diabetes educator I knew this could become a serious problem. Soon after my husband started taking prednisone, his blood glucose level went over 200. Then he started taking nopal cactus, and it dropped to 150 and then to 132. We're pleased with the results but would like to know more about nopal.

A People have eaten prickly pear cactus leaves (*Opuntia*) as a vegetable for centuries. Nopal, as it is known in Mexico, has also been used there to lower blood sugar in type 2 diabetics. Some preliminary animal research suggests that nopal is effective, but there is little human data to support its use. Anyone who might consider such an approach needs to be under medical supervision and monitor blood glucose carefully. Nopal capsules can be found in health food stores or on the Internet.

NUTS

Q Could you talk about the relative benefits of various nuts? We know almonds and walnuts can lower cholesterol. Do pecans, macadamia nuts, or hazelnuts offer anything besides calories?

A Nearly all nuts have beneficial fatty acids, particularly monounsaturated fatty acids like those found in olive oil. In addition, walnuts contain some omega-3 fatty acids like those found in fish oil. People who eat five ounces of nuts weekly are less susceptible to heart disease and type 2 diabetes. The nuts must be part of the diet, though, and not added to it. Excess weight from added calories can raise the risk of developing type 2 diabetes.

VINEGAR

Q I suffer from type 2 diabetes. My doctor prescribed Glucotrol for my blood sugar. It helped to a degree, but I have found that by adding apple cider vinegar and cinnamon to a careful diet, I can control my blood sugar even better. I know I haven't made this up, but are there any data showing that these natural remedies work?

High-carbohydrate meals containing white bread or rice can raise blood sugar. We would not have imagined that vinegar could counteract this effect, but there is growing evidence to support your experience. Scientists in Sweden report that vinegar given with white bread reduces blood sugar and insulin. It also helps people feel full up to two hours later.[3] Japanese researchers have found that vinegar can counteract the effect of white rice on blood sugar. And investigators at Arizona State University report that two tablespoons of vinegar before a starchy meal can significantly reduce the expected rise in blood glucose.[4]

Diarrhea

Diarrhea can be caused by an infection, as traveler's diarrhea generally is. Prudent eating and drinking (and daily Pepto-Bismol) can sometimes prevent it, but once the symptoms start, it is too late for prevention. Antibiotic medications can help. People with chronic diarrhea due to other causes frequently suffer for years. Sulfite sensitivity can trigger a bad bout. Irritable bowel syndrome (IBS) can also be troublesome. Here are some of the home remedies our readers and listeners have found helpful.

COCONUT

I'd like to try coconut macaroons for controlling diarrhea, but I am trying to cut out sugar. Could I make my own macaroons using Splenda instead of sugar? Can I use shredded coconut by itself instead of eating the cookies?

A Readers tell us that eating two coconut macaroons daily can ease chronic diarrhea. Some report that plain coconut also does the trick.

Q I noticed a report on your website (*www.peoplespharmacy.com*) about coconut macaroon cookies stopping diarrhea. I am a hospice nurse, and one of my patients was literally dying of diarrhea. None of the medical treatments were helping. After reading about macaroons, I asked my patient's daughter if we could try giving her mom coconut milk and rice milk. She had nothing to lose. The diarrhea stopped in 24 hours, and the patient began to eat again. As a result, she began to thrive and had to leave our hospice program!

A Flunking out of hospice is good news. We are so glad this suggestion inspired you and was so helpful for your patient with life-threatening diarrhea. We have heard from many readers that coconut is a traditional remedy for diarrhea.

Q Thank you so much for writing about IBS and coconut macaroon cookies. It works. I suffered with chronic diarrhea for years and have been healed for the last two years.

A We're always pleased to hear about success with home remedies. Donald Agar wrote us nearly ten years ago to report that two Archway Coconut Macaroon cookies a day banished the chronic diarrhea he suffered as a consequence of Crohn's disease. We have heard from other readers that coconut helps combat diarrhea. But not everyone who suffers with IBS benefits from coconut macaroons.

Q Thank you for your invaluable advice. After a colon resection I had severe diarrhea diagnosed as IBS. Nothing helped except four to six Lomotil pills daily. Then I tried your remedy of shredded coconut, and it worked like a miracle. Can I safely increase the dosage, which is presently three teaspoons?

A It is possible to increase amount of coconut you are taking. Be careful, though, since too much may lead to constipation. Coconut is high in fat and calories. So you may need to adjust your diet to compensate. See recipe for Coconut Chicken Soup in the Colds entry, p. 51.

Q I have a student with irritable bowel syndrome. Nothing—even removing her gallbladder—gave her relief. She was thinking of dropping out of school. I told her what I read in your column about coconut, and it worked. Her digestion is normal again. Amazing!

A Some people report that two Archway Coconut Macaroon cookies daily control chronic diarrhea.

DIETARY DISCRETION

Q Some of your readers have had questions about chronic diarrhea. My mother had a similar problem for many years. Then one day her doctor suggested that it could be caused by lactose intolerance. He hit the nail on the head. My mother loved ice cream, and the milk sugar in it was triggering the diarrhea.

A Lactose, the sugar in milk, is indigestible for many adults who lack an enzyme called lactase. For such people, drinking milk or eating ice cream or other dairy products can cause gas, bloating, cramping, and diarrhea. The safest way to prevent symptoms is to avoid all forms of milk and milk sugar (which is sometimes used in prepared foods or even as binders in pills). Commercial lactase, such as Lactaid, Lactrase, and generic pills, can sometimes be helpful. There are even dairy products that have been pretreated to reduce lactose.

Q Do you have any information regarding sugar-free gum and diarrhea? My daughter had trouble with weight loss, stomach cramps, and diarrhea. Three different doctors could not diagnose the cause. Then she remembered it all started after she began chewing sugar-free gum.

A Sugar-free gum frequently contains compounds such as maltitol, sorbitol, mannitol, and xylitol. These sweeteners are not absorbed well from the digestive tract, and they attract water. This can lead to watery diarrhea, gas, and cramps. Giving up sugar-free gum should ease your daughter's digestive woes.

Q My husband is plagued with diarrhea. He'll be okay for a week or so. Then, for no apparent reason, he has diarrhea. He's been eating two coconut macaroons a day for about two weeks. We thought that had taken care of the problem, but it appeared again today. I read that sugarless gum could cause diarrhea. He chews it every day. Can you tell me about this?

A Many readers report that eating two coconut macaroon cookies daily helps control their chronic diarrhea. But why treat a problem that might be avoided? Sugarless gum could be the culprit in your husband's case. Sweeteners in sugarless gum, such as sorbitol, maltitol, mannitol, and xylitol, are poorly absorbed from the digestive tract. When the residue reaches the large intestine, it can cause gas and diarrhea. Your husband should try giving up the gum to see if that solves the problem.

FISH OIL

Q Has anything shown that fish oil is effective in reducing chronic diarrhea? It worked instantly for me, although I started taking it for heart health. The results were startling—no more runs or trots.

A The only research we found on fish oil for diarrhea involved studies on rats.[1] Scientists gave rats a drug that caused chronic diarrhea, and fish oil sped intestinal repair and recovery. We don't know if this will work for anyone else, but fish oil has enough health benefits to be well worth a try.

POMEGRANATE JUICE

Q Some of your readers have asked about chronic diarrhea. One of the best things to take is pomegranate juice, which can be found in grocery stores. You can actually get constipated if you drink too much (as I found out!).

A Researchers are rediscovering the healing power of pomegranates. Studies suggest that this ancient fruit may help reduce the risk of blood clots and keep cholesterol from damaging arteries. People have traditionally used pomegranate to treat diarrhea and dysentery, though there isn't any research to show it is effective. We'll go along with the warning not to overdo.

PROBIOTIC YOGURT

Q I had chronic diarrhea for several years, so I was interested in the coconut macaroon cookie remedy when I read about it in your column. I ate two each morning and got a benefit for a while, but then I had to increase the dose. After a few months, even three cookies were not helping the diarrhea. Instead I turned to Dannon Activia yogurt. They advertise that they will refund your money if their product doesn't solve the problem in two weeks. I didn't get any money back, but I am happy. It not only eliminated my diarrhea, but also solved my husband's long-standing constipation problem.

A Activia contains probiotic bacteria that are supposed to help reestablish a healthy balance of microbes in the gut. Yogurt is made from cultured milk, so it is an excellent way to deliver living bacteria. Probiotics have gained popularity in Europe but are still relatively unknown in the United States. Nonetheless, some research links probiotics to improved digestive health.

QUESTRAN

Q One of your readers was troubled with diarrhea. After gallbladder surgery, I had the same problem. Saying the bile acid had caused the diarrhea, my doctor prescribed Questran. It works great.

A Cholestyramine (Questran) lowers cholesterol by binding to bile acids. Along with relieving diarrhea, it keeps blood lipids under control.

THE PEOPLE'S PHARMACY

Favorite Food #9: Yogurt

Over the last few years, the term "probiotics" and the phrase "live and active cultures" have gone mainstream. We've been excited about probiotics for years, and we're glad that large yogurt manufacturers, like Dannon, have taken up the cause of this duly celebrated supplement. Dannon's Activia contains probiotic bacteria that repopulate the stomach and gastrointestinal tract with good bacteria and push bad bacteria out. Because yogurt is made from cultured milk, it is a very effective way to deliver living bacteria.

Research suggests that this rebalancing should soothe a vengeful gut and ameliorate conditions from gas to diarrhea to diverticulitis.[1] There is also some research indicating that probiotics may benefit the immune system in other ways—easing eczema symptoms and even helping reduce susceptibility to upper respiratory tract infections.

Yogurt offers even more advantages. For one, calcium, when ingested in the form of food (rather than pills), may protect people from kidney stones. And many women have long known that eating yogurt can help prevent yeast infections. In one study, women who ate a cup of yogurt containing live *Lactobacillus acidophilus* cultures each day were three times less likely to have a recurring yeast infection.[2]

Not all yogurts are created equal, however. Some yogurts that may taste great are just high-sugar snacks that should be treated as dessert rather than health food. Look for a label that says "probiotics" or "live and active cultures" (the National Yogurt Association has developed an official seal). But keep in mind that calcium can interfere with the absorption of some medications, like certain antibiotics and osteoporosis drugs. Talk to your doctor or pharmacist about whether eating yogurt could affect any of the drugs you're taking, and if so, how long you should wait to eat yogurt after taking them.

Diverticulitis

This digestive condition is another mystery. Why do some people develop little pouches (diverticula) along the wall of their colons? Why do some people develop inflamed or infected diverticula (diverticulitis)? There are lots of theories, but the predominant hypothesis is lack of fiber (roughage). It might even be true. But a fair number of folks eat lots of fiber and fruits and vegetables and still end up with diverticulitis. And there are those who eat very little roughage and escape unscathed.

When we don't know for sure, we frequently blame our genes, and they too may be a contributing factor. Regardless of what causes diverticulitis, it can be painful and disabling. Symptoms can include abdominal pain, fever, nausea, cramping, bloating, diarrhea, and constipation. Treatment often includes antibiotics if there are signs of infection within the diverticula. Not infrequently, physicians will give the bowels a "rest." If patients are hospitalized, doctors may take them off food and give them only intravenous fluids for a while. Outpatients often receive the confusing advice to eat a very low-fiber diet initially during an attack. Then, eventually, when the colon recovers, they are encouraged to consume a high-fiber diet.

Diverticulitis requires excellent supervision by a very knowledgeable gastroenterologist. If the situation gets serious, surgery may be necessary. We hope these suggestions help prevent that.

PROBIOTICS

Q My husband was diagnosed with diverticulitis. He was treated with antibiotics, but the doctor said he could have another attack at any time. My husband now avoids seeds and nuts, but a different doctor says food has very little impact. I now give my husband lots of fruit, yogurt, and acidophilus milk, and he is taking FiberCon daily. Is there anything else that might help?

A Your husband may want to try probiotics (good bacteria). Such products are available in health food stores and in many supermarkets. One reader reported, "After ten years of being diagnosed repeatedly with diverticulitis and treated with antibiotics, my digestive system went crazy and I lost bowel control. More antibiotics and prednisone were prescribed. One doctor wanted to do surgery, perhaps a colostomy. I sought a second opinion, and the doctor prescribed probiotics. A week later I was fine. After four years I have no more diverticulitis and no diarrhea."

Dry Skin

Dry skin can be terribly annoying. Some things seem to make it worse: frequent hand washing, exposure to water or soap, winter weather with wind and dry indoor air, and health conditions like eczema or an underactive thyroid gland. No matter what, you have to wash your hands to stay healthy, but using rubber gloves to do dishes or other cleaning chores may help. Dermatologists also suggest avoiding soap—instead, use a waterless cleanser for hands or a soap-free cleanser such as CeraVe or Cetaphil for the face or body. Using a good moisturizer after gently patting the skin dry is essential.

At night, try wearing a relatively greasy moisturizer, such as Aquaphor Healing Ointment or Vaseline Petroleum Jelly—under cotton gloves to protect bedclothes. Some of our other favorite remedies are listed below.

BARNYARD BEAUTY AIDS

Q My late husband always made sure we had Bag Balm in the house to use for scrapes and cuts. Now our children keep it in their homes, and our granddaughter uses it daily. I first started applying it to some horrible big cracks on my heels. It worked so well that I now use it on my arms, legs, face, and neck. It is better than any other lotion for my dry skin. I mix it with a little cold cream so it goes on more smoothly, and it sure makes a huge difference.

A Thanks for sharing your skin care tip with us. Many readers appreciate the moisturizing power of beauty aids like Bag Balm and Udderly Smooth Udder Cream.

Q I get dry hands every winter. Cracks in my fingertips and knuckles drive me crazy. My nails are rough too, and I am at my wit's end. I cannot afford costly department store products. Any suggestions?

A In winter we frequently recommend "barnyard" beauty aids. Dairy farmers learned long ago that the salves they used to prevent cows' udders from chapping also worked beautifully for their own hands. The oldest is Bag Balm from the Dairy Association Company (800-232-3610). It is greasy and smelly, though. Udderly Smooth Udder Cream is nicer to use and also provides good moisturizing at a good price (800-345-7339). Listeners will recognize Udder Cream (with urea, for extra moisturizing power) as a longtime underwriter of the People's Pharmacy radio show.

Q My skin is so dry and itchy that sometimes I have to stop and scratch it in the middle of a tennis game. This annoys my partner and bothers me more than I can say. What suggestions do you have?

A For very dry skin, dermatologists often recommend a heavy-duty moisturizer. Plain petroleum jelly works well and is inexpensive. Aquaphor Healing Ointment is also very effective, though some people find it greasy. The best way to use such products is by applying them after a bath. Blot (don't rub) excess water off the skin, and then put the moisturizer on. Many people also find "barnyard" moisturizers such as Bag Balm or Udderly Smooth Udder Cream helpful and cost-effective.

OLIVE OIL

Q Years ago my dermatologist suggested I stop using topical creams and lotions since I am allergic to them all. Twice a day I rub olive oil on my skin instead. On weekends I also use it as a hair conditioner. By sticking with olive oil, I have solved my skin problems. As the doctor said, "If it was good enough for Cleopatra, it's good enough for you!"

A Some people may be allergic to olive oil, but for most people it can often be an effective moisturizer. It may be a little greasy, however. Some women also find that applying olive oil can help reduce vaginal dryness.

VINEGAR

Q I used to have severely dry skin. My hands were always dry and chapped. Then a friend told me to dip my hands in a solution of two-thirds white vinegar and one-third water for one or two minutes and then rinse them off. I keep a spray bottle of that mix in my shower to spray on my feet and hands. My heels are no longer so dry and rough that they tear my hose. I have given this tip to hairdressers who have dry hands because of the chemicals they use.

A This is not the first time we have heard that vinegar could help dry skin. There are no scientific studies to support this claim, but this inexpensive remedy may be worth a try. One theory is that vinegar restores balance to dry skin caused by too much hand washing.

Eczema

No one knows exactly what causes the red, itchy skin condition called eczema. But we do know a few things that exacerbate eczema: stress, irritants like laundry soap, prolonged exposure to water, and contact with allergens. The itching caused by severe eczema can be extremely distressing. Doctors may prescribe heavy-duty immune-suppressing creams such as Elidel or Protopic, but worries about the risks posed by these drugs have led many people to seek home remedies.

BORAGE OIL

Q I am a 51-year-old female plagued with persistent eczema. The skin on my hands was always red, itchy, cracked, and often bleeding. My hands were always covered with bandages or gauze. Dermatologists prescribed cortisone creams of increasing strength. None of them were helpful over the long term. Hand cream for dry skin was totally useless. Five years ago I went to an allergist for an unrelated problem. When he saw my hands, he was concerned that the open sores put me at risk of infection. He suggested taking borage oil. I tried taking one capsule of borage oil after breakfast and one before bed. Within a few months the eczema on my hands had disappeared completely, and the condition is now only a minor annoyance. I control my dry skin with ordinary hand cream. I hope this tip will help others.

A Borage oil is rich in a fatty acid called gamma-linolenic acid (GLA). The oil comes from the plant *Borago officinalis,* also known as starflower. We are delighted that you got such relief, but not everyone will benefit. A placebo-controlled study suggests that borage oil is ineffective for eczema. The researchers conclude, "It seems unlikely that dietary supplementation with gamma linolenic acid is beneficial in management of atopic dermatitis."[1]

CERAVE MOISTURIZER

Q I have had eczema ever since I was a child. I have used many steroid creams over the years, and while they help a bit, in bad bouts the creams were not soothing and just kept away the worst irritations. I have been going to a young dermatologist who advised me to use CeraVe Moisturizing Cream (not lotion). I can't rave about it enough. Immediately after bathing I put it on, and 24 hours later, when I shower again, I can feel that the cream is still there. I have only had to use one prescription cream a few times since starting with this more than a year ago. Keeping my skin hydrated seems to do the trick for me.

A CeraVe moisturizer contains no fragrance to irritate the skin, but it does contain ceramides. These are natural fatty compounds found in cell membranes. People with eczema frequently have abnormally low levels of ceramides in their skin. Moisturizing can help keep eczema from itching and may boost the effectiveness of topical steroids when you do need to use them. Another interesting product is CamoCare Soothing Cream (*www.camocare.com*). It contains a chamomile-derived oil that has anti-inflammatory properties.

DERMASMART CLOTHING

Q My eight-year-old son has eczema. We have been alarmed by the recent studies about Elidel increasing the risk of cancer. We also do not want to go back to topical steroids because they might thin his skin too much. Are there any other treatments that we can consider?

A Besides using a good moisturizer to keep the skin from drying out, you may want to consider DermaSmart undergarments and pajamas (*www.dermasmart.com*). This special fabric is supersoft and nonirritating. A firefighter told us that he developed eczema after exposure to mold in an older fire station. When he put on his protective gear the itching nearly drove him crazy. The DermaSmart T-shirt and pants reduced the irritation and itching.

INSTANT GLUE

Q Like some of your readers, I get cracked fingers in cold weather. This year I decided to try something different. I had a place that split open on my thumb. I put Super Glue on it, held it together, and let it dry. It worked beautifully. Instant glue seals the crack so there's no pain. When it starts washing off, I just put more on. Works great!

A Physicians have been using a product similar to instant glue to close cuts. The cyanoacrylate glue they use is called Dermabond, which is now also available to consumers in Band-Aid Brand Liquid Bandage. It may be less irritating to the skin than household instant adhesives like Super Glue or Krazy Glue.

LOW-CARB DIET

Q I used to get urinary tract infections or yeast infections every other month. Then I changed my diet. I cut out sugar, white flour, and starches like potatoes and rice. Since then I have had only one urinary tract infection. I've lost 20 pounds, and my eczema is 99 percent better. I only have a flare-up when I have cake or milk chocolate. I am still surprised that diet can have such an effect. Other people with eczema or seborrheic dermatitis might benefit the way I did.

A There is not much research linking a high-carb diet to urinary tract infections or eczema. But reducing the amount of sugar, starch, and refined carbohydrates seems like a simple enough experiment. If it works for some people with such hard-to-treat conditions, it might be worth the trouble. Thanks for your interesting story.

NOXZEMA

Q I just had to let you know the success I've had with your suggestion to use original Noxzema for eczema. My three-year-old son has suffered from this skin condition on his legs and feet for two years. We treated it successfully with the prescription drug Elidel, but safety concerns led us to stop using it. I tried many moisturizing creams to soothe his skin, but he cried and said they hurt. I started using Noxzema the day I read your article, and there were no tears. His skin responded quickly, and after three weeks almost all traces of eczema are gone. This advice has changed my son's life.

A We are pleased to learn of your success. Lore has it that Noxzema was named after the product helped a customer "knock" eczema.

OOLONG TEA

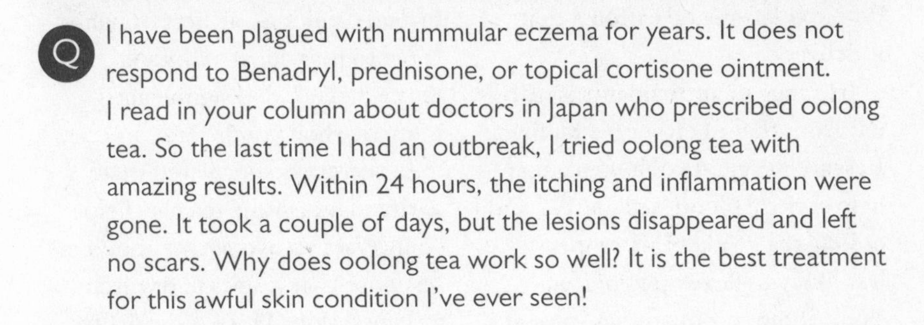

Q I have been plagued with nummular eczema for years. It does not respond to Benadryl, prednisone, or topical cortisone ointment. I read in your column about doctors in Japan who prescribed oolong tea. So the last time I had an outbreak, I tried oolong tea with amazing results. Within 24 hours, the itching and inflammation were gone. It took a couple of days, but the lesions disappeared and left no scars. Why does oolong tea work so well? It is the best treatment for this awful skin condition I've ever seen!

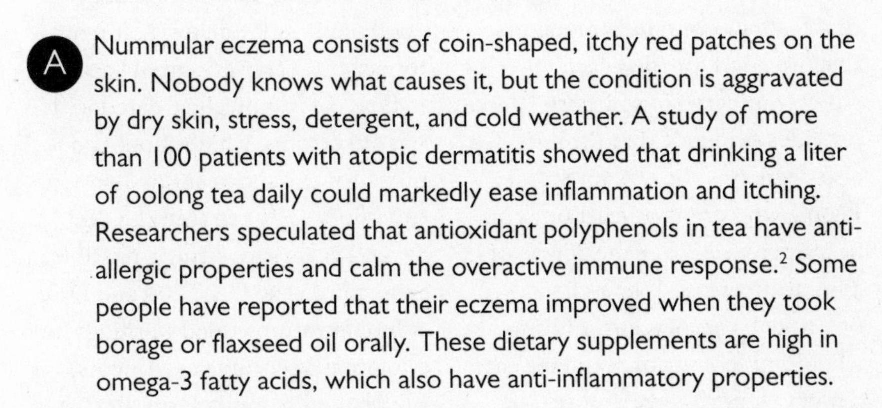

A Nummular eczema consists of coin-shaped, itchy red patches on the skin. Nobody knows what causes it, but the condition is aggravated by dry skin, stress, detergent, and cold weather. A study of more than 100 patients with atopic dermatitis showed that drinking a liter of oolong tea daily could markedly ease inflammation and itching. Researchers speculated that antioxidant polyphenols in tea have anti-allergic properties and calm the overactive immune response.[2] Some people have reported that their eczema improved when they took borage or flaxseed oil orally. These dietary supplements are high in omega-3 fatty acids, which also have anti-inflammatory properties.

PROBIOTICS

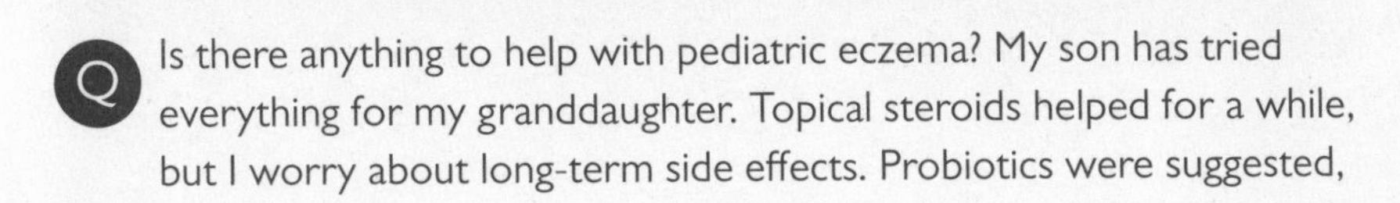

Q Is there anything to help with pediatric eczema? My son has tried everything for my granddaughter. Topical steroids helped for a while, but I worry about long-term side effects. Probiotics were suggested,

THE PEOPLE'S PHARMACY

Favorite Food #10: Oolong Tea

Oolong, whose name is nearly as delicious as its taste, is a traditional Chinese tea. It belongs to neither the black nor the green tea families and falls somewhere between them in the amounts of both oxidation and caffeine that it delivers.

Its range of antioxidants is different from that of either tea family. Research suggests oolong tea may help control blood sugar levels, and oolong tea drinkers also appear less likely to develop hypertension.[1] Some recent studies suggest that tea may contain compounds that are good for the heart; one study conducted in southern China provides evidence that drinking tea also decreases the risk for stroke.[2] People who drank at least one cup of tea per week over the course of 30 years decreased their chances of stroke by 60 percent. Those who drank oolong or green tea each day cut their risk even more—by more than 70 percent.

But there's more. Oolong tea has one benefit that no other hot beverage can claim: It is remarkably effective at conquering the pesky, scaly skin condition eczema.

More than 100 people suffering from different types of eczema participated in a study that confirmed drinking one liter (roughly three to four cups) of oolong tea each day could significantly improve their condition.[3]

Researchers speculated that antioxidants in the tea may help counteract an overactive immune response that results in dry, itchy, inflamed skin. This is especially good news, since many treatments for eczema carry potential hazards.

Eczema is typically treated with steroid creams, which should be taken for the shortest possible time since they can thin the skin. Several years ago, citing a possible cancer link, the U.S. Food and Drug Administration issued warnings for two popular eczema treatments, Elidel and Protopic. Drinking tea, steaming hot in winter or iced in summer, poses none of these potential risks—and it's tasty!

but I don't know anything about them. Any information would be most welcome.

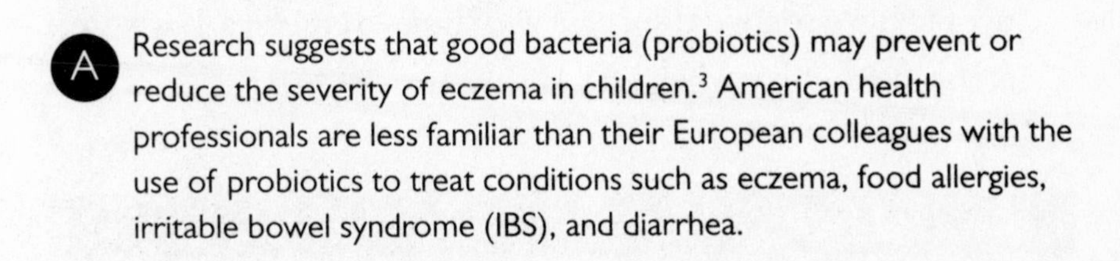

A Research suggests that good bacteria (probiotics) may prevent or reduce the severity of eczema in children.[3] American health professionals are less familiar than their European colleagues with the use of probiotics to treat conditions such as eczema, food allergies, irritable bowel syndrome (IBS), and diarrhea.

PYCNOGENOL

Q For many years I have suffered from atopic dermatitis, a type of eczema. The regimen prescribed by my dermatologist helped to some degree, but what seems to have cleared up the itching is an antioxidant supplement called Pycnogenol. After six months, my eczema is in total remission. What can you tell me about Pycnogenol?

A Pycnogenol is extracted from the bark of maritime pine trees. The plant compounds it contains are proanthocyanidin flavonoids similar to compounds derived from cranberries, blueberries, and other plant sources. Pycnogenol has antioxidant and anti-inflammatory properties. Perhaps the effect on your eczema is related to its ability to reduce inflammation.

Fibromyalgia

This mysterious condition causes repeated bouts of soft tissue pain, weakness, and fatigue. People with fibromyalgia often have disturbed sleep and may wake up tired even after eight hours in bed. Diagnosis of fibromyalgia is difficult, in part because many doctors believe it is a psychosomatic condition or a form of depression. Very low-dose antidepressants at bedtime may relieve pain or help with sleep. Gentle exercise can also be helpful. Vitamin D is also said to ease the discomfort of fibromyalgia. Relatively few people have reported success with home remedies, but here are some that have helped.

BOSWELLIA

Q I suffer from fibromyalgia, which is an extremely painful illness. Lately I have found that the herb boswellia is helpful for the morning stiffness that goes with fibromyalgia. There are no side effects, and the results are very quick. It doesn't eliminate pain, but reducing my stiffness makes the morning less difficult. Others with fibromyalgia might want to know about this.

A Thank you for sharing your experience. Boswellia is an herb used in the ayurvedic tradition, an ancient medical system of India that makes use of plants. The herb has anti-inflammatory properties, which might explain why it helps reduce stiffness associated with this debilitating chronic condition. Most studies have found no significant side effects.

CHERRIES

Q I have spinal cerebellar ataxia, a genetic disorder characterized by lack of coordination and a halting gait, and I also have fibromyalgia. Tart cherry concentrate is the greatest thing for easing my discomfort. I don't even take pain pills anymore.

A Others have told us that tart cherry, as juice or concentrate, is helpful against the pain of gout or arthritis. We are glad to learn it has also helped your fibromyalgia.

Gas

Flatulence isn't harmful, but it can be embarrassing. Everybody has gas, but some people seem to have more gas than others. Diet or even medications may be to blame, but home remedies sometimes help.

BITTERS

Q I would like to know the formula you once printed about using bitters for flatulence.

A The Angostura bitters label suggests taking one to four teaspoonfuls after meals for flatulence. Some readers put it in club soda or 7UP to mask the bitter taste.

FENNEL TEA

Q For the past several months I have been suffering with flatulence. It is extremely embarrassing. I worry every time I go out in public that I will pass smelly gas. I have tried over-the-counter medications like Gas-X, Beano, Tums, and charcoal capsules. I try to avoid foods that might give me gas, but even so, the problem persists. Is there anything I can do? It is getting so bad that I don't want to go out in public anymore.

A It sounds as if you have tried almost everything in the pharmacy. Fennel tea is a home remedy to consider. Crush one teaspoon of fennel seeds with a spoon and steep it in hot water for five minutes. Some flatulence sufferers have also suggested using a dose of Pepto-Bismol to help control odor.

THE PEOPLE'S PHARMACY

Favorite Food #11: Fennel Seed

This member of the celery family is a well-known herb native to southern Europe and western Asia, but it was also known in ancient China as well as in India, Egypt, and Greece. In the Middle Ages people prized it as a vegetable, and indeed we appreciate its flavor today. Colonists brought fennel seeds to the New World.

This aromatic little spice might seem modest and mild. But in reality, it has the power to tame unpleasant flatulence and heartburn. For many years, we have heard from readers who have banished gas by sipping fennel seed tea. Some report that it's also possible to get relief by chewing on the actual seeds. In fact, many Indian restaurants offer fennel seeds as a colorful after-meal treat to aid digestion. They can also be purchased at health food stores.

For people who don't like the licorice flavor, fennel seeds are available in capsule form as well. (In fact, fennel is not actually related to licorice and therefore carries none of its potential dangers, like increased blood pressure, fluid retention, depleted potassium, and disrupted hormonal balance.) Anyone allergic to celery, carrots, dill, or anise should avoid fennel. Pregnant women should not use fennel oil or extracts, but fennel seed infusions are probably safe.

We have heard that, in addition to soothing the stomach, fennel seeds may ease upper respiratory and sinus symptoms. One reader uses this recipe for sinus trouble:

1 tablespoon fennel seeds
¼ teaspoon powdered ginger
1 clove
½ inch piece of cinnamon stick
1 teaspoon brown sugar

Combine ingredients in two cups of water. Boil until one and a half cups of liquid is left. Strain mixture and drink it hot with milk. You can substitute honey for brown sugar.

Fennel Seed Tea

½ teaspoon fennel seeds
1 cup boiling water

Smash seeds with a spoon to "bruise" them. Steep in water for five minutes. Strain out seeds and sweeten to taste.

FLATULENCE CUSHION

Q I am interested in purchasing flatulence filters. My mom has severe gas problems and is taking charcoal tablets along with other things. The problem is still noticeable. I would like to try the filters. I have checked drugstores but have had no luck.

A Search the Web for the GasBGon flatulence filter seat cushion (*www.gasbgon.com*). It contains activated charcoal that traps odors from the digestive tract. The same company (Dairiair at 877-427-2466) also makes underwear with activated charcoal woven into the fabric. These carbonized undies reduce unpleasant smells even when the wearer is not sitting on a flatulence filter seat cushion.

PEPPERMINT OIL

Q Your newsletter had a story about a lady who needs help for the gas caused by her mother-in-law's cooking with onions, cabbage, beans, and barley. I've found peppermint oil capsules (sometimes sold as breath fresheners) are brilliant for quick relief. Prevention would be better, however. My grandmother always boiled her onions first and then strained the water off before adding them to her recipe. This removed the gas-causing part, and we could all enjoy eating her meals with no worries.

A Thanks for the recommendation of peppermint oil capsules. Some readers claim that mint tea drunk after a meal can ease gas. Your suggestion of discarding the initial cooking water can help with beans as well as onions.

PROBIOTICS

Q What is the best thing to do for daily, embarrassing gas? I eat high-fiber foods during the week, but more of a variety on weekends, and it doesn't seem to make a difference. I have continuous gas all week.

A record of what you eat and how your gut responds may be helpful in pinpointing whether you are reacting to a specific food. Some people find that milk and dairy products cause distress. Others have trouble with foods like bagels or pretzels. Once you identify a likely culprit, avoiding it should tell you if you were correct. Another reader found a simple solution: "Activia yogurt is excellent in stopping flatus. One small carton a day stopped most of the problem." Activia contains probiotics (good germs) that can aid digestion.

Gout

Anyone who has suffered from it can tell you that gout can be excruciating. In this condition, uric acid crystals precipitate out of the blood into the joints. The joint of the big toe is especially vulnerable to gout, although the condition can also affect ankles, knees, elbows, and wrists. During an attack, even the light pressure of a bedsheet may be unbearable. Doctors prescribe a number of medications for gout, and one or more may be necessary. When home remedies work, however, they are less expensive and less likely to cause side effects.

CELERY

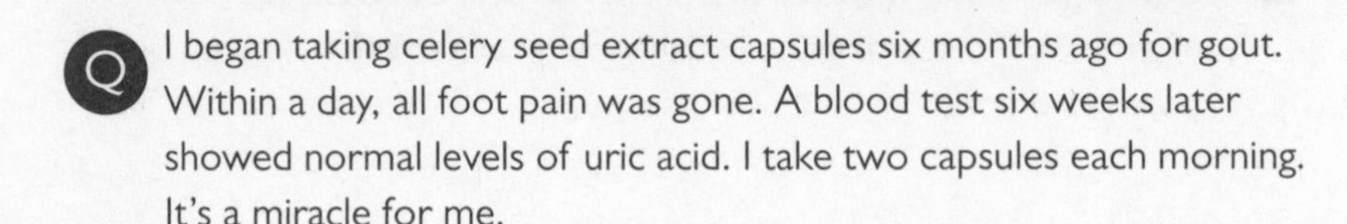

Q I began taking celery seed extract capsules six months ago for gout. Within a day, all foot pain was gone. A blood test six weeks later showed normal levels of uric acid. I take two capsules each morning. It's a miracle for me.

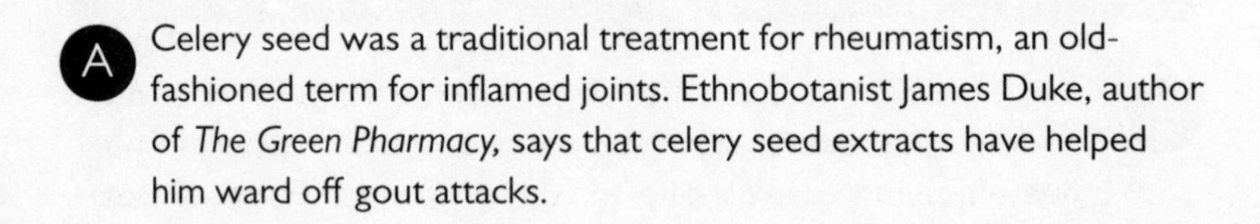

A Celery seed was a traditional treatment for rheumatism, an old-fashioned term for inflamed joints. Ethnobotanist James Duke, author of *The Green Pharmacy,* says that celery seed extracts have helped him ward off gout attacks.

CURRY (TURMERIC)

Q I suffer from gout from time to time. Have you ever heard of curry relieving the symptoms? My wife made a soup containing curry powder, and within one hour after eating it, I could feel the pain going away. I ate the soup the next two nights, and my gout was 95 percent gone.

A Thanks for the tip. Gout is a painful inflammatory condition in which uric acid crystals collect in the joints. The yellow spice in curry powder is turmeric. For centuries ayurvedic practitioners have used it to treat inflammation. Research on animals confirms that turmeric extracts can reduce joint swelling from arthritis.[1]

SOUR CHERRIES

Q I have heard that sour cherry juice can ward off gout attacks. Have you heard of this remedy? My doctor thinks it is ridiculous.

A Cherries have traditionally been recommended for gout prevention, but the medical evidence has been limited. One study showed that uric acid drops after people eat bing cherries.[2] Elevated uric acid triggers the excruciating pain of a gout attack, so this finding supports the potential usefulness of sour cherries against gout.

Q My sister has had two recent episodes of gout. Without health insurance, she could not afford to go to the doctor. I gave her some samples of an anti-inflammatory medicine I had on hand. She took them but got relief only when she started eating sour cherries. Someone told her it was an old remedy to eat six cherries a day. Was this relief all in her mind? Will the gout return? It was extremely painful and left her almost immobile.

A Once someone has an attack or two, they may be more prone to others unless their uric acid levels are lowered. Many readers have reported that sour cherries can ease the pain associated with gout and even arthritis. Fresh, dried, or frozen cherries, cherry juice, or even cherry extract capsules may be helpful. No home remedies have been clinically tested, however. There are medicines that help lower uric acid levels, but they require a prescription. A diet low in red meat, fish, and other seafood and high in low-fat dairy products seems to help some people avoid gout attacks. One study confirms that alcohol, especially beer, increases the risk of gout.[3]

Q I was on Celebrex but experienced side effects. A friend recommended that I try FruitFast Tart Cherry Juice Concentrate from Brownwood Acres. It took four weeks for the juice to kick in, but at the ripe old age of 79 I'm tap dancing again. It worked for me.

A We've heard from others that tart or sour cherries or cherry juice might ease joint pain from gout. Your testimonial is terrific, and we suspect others will want to try cherry juice for arthritis as well. The brand you mention is available at 877-591-3101 or *www.brownwoodacres.com.*

Q I was fascinated to read about a 79-year-old person with arthritis who is tap dancing again after drinking tart cherry juice. I too have been drinking the juice in addition to watching my diet and cutting out shellfish. I have had no further attacks.

A We've received many anecdotal reports that tart cherries help relieve gout. You are smart to watch your diet. One article confirms that people who eat a lot of meat and seafood are more inclined to have gout attacks.[4]

Headaches and Migraines

We take head pain very seriously. If you are experiencing chronic or severe headaches, you should waste no time getting to a doctor or headache specialist. Pain in the head can signal a variety of extremely serious conditions. That being said, headaches are a common affliction, and, statistically speaking, far more often than not they're simply a painful nuisance rather than a medical emergency.

The first thing to know about run-of-the-mill headaches is that medicine taken to alleviate pain might actually be causing the symptoms. Medication-overuse headaches often occur in people who pop painkillers two or more times a week, as their brain receptors become sensitized and trigger more and more severe headaches. The way to break out of the cycle is to stop taking the medication (under a physician's supervision) and to go through an adjustment period. Lots of other things can cause headaches as well. For example, gluten can trigger headaches in people suffering from celiac disease. Forgoing that morning cup of joe can literally be painful for folks addicted to caffeine. And as anyone who's ever had a hangover knows, overindulging in alcohol can lead to a rude awakening the next morning.

BEER

Q I know that red wine can cause migraines. But I have also heard that there is something you can drink to alleviate a migraine. Do you have any idea what it could be?

A This will sound bizarre, but one person told us that drinking a beer at the very first sign of a migraine could prevent the attack from progressing. She learned this approach from an old country doctor and has used it successfully for decades. We can't explain why beer might work against migraines, but we have heard from others that it can be helpful.

DIETARY TRIGGERS

Q I suffered from migraines for years. Then I started taking magnesium and practicing yoga. I also eat a low-carbohydrate diet—low in yeast, sugars, and grains. I no longer suffer. It's a blessing.

A We're glad to hear these changes made such a difference for you. We've also heard from those who have changed their diets in other ways, including adding supplements like magnesium, vitamin B complex and B_2 (riboflavin), coenzyme Q10, melatonin, feverfew, and butterbur.

Q I suffer from classic migraines. I've been able to cut way down on their frequency by avoiding my food and beverage triggers and taking blood thinners such as aspirin, vitamin E, and omega-3 and omega-9 fatty acids. This has stopped my migraines almost completely.

A There are, unfortunately, a lot of foods and drinks—such as caffeine, chocolate, red wine, aged cheese, or too much of any kind of alcoholic beverage—that may trigger migraines in some unlucky folks. Frequent culprits also include foods containing aspartame, MSG, sulfites, nitrates, and tyramine. But there are also foods and beverages that may help alleviate headache or migraine symptoms.

GINGER

Q I sometimes make a tea of mint, chamomile, sassafras (which one grandmother called "headache bark"), some cinnamon sticks, cloves, and a bit of valerian. I add grated fresh ginger when preparing the tea. It's not a cure, but it helps, as do ginger beer; a warmed, buckwheat-filled pack along my back and shoulders; and ice packs on my temples and forehead. (I wish Imitrex and its relatives worked for me, but they don't.)

A Studies have documented ginger as a migraine treatment for decades.[1] A small study testing a combination product (GelStat Migraine) containing ginger and the herb feverfew indicated that the product could help alleviate migraines.[2]

GLUTEN-FREE DIET

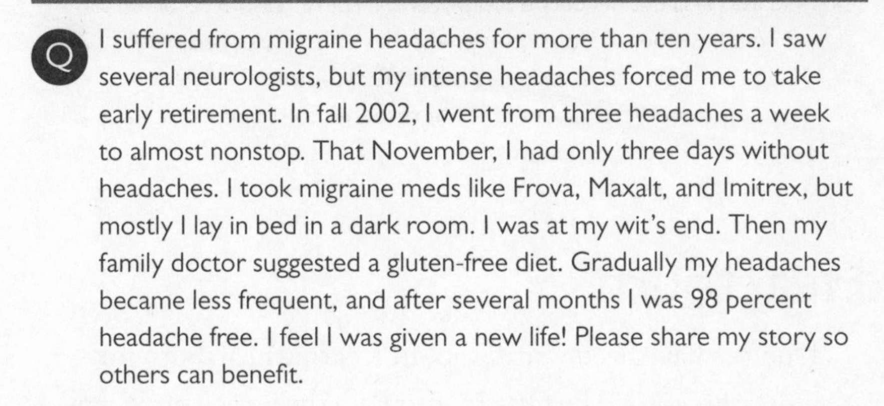

Q I suffered from migraine headaches for more than ten years. I saw several neurologists, but my intense headaches forced me to take early retirement. In fall 2002, I went from three headaches a week to almost nonstop. That November, I had only three days without headaches. I took migraine meds like Frova, Maxalt, and Imitrex, but mostly I lay in bed in a dark room. I was at my wit's end. Then my family doctor suggested a gluten-free diet. Gradually my headaches became less frequent, and after several months I was 98 percent headache free. I feel I was given a new life! Please share my story so others can benefit.

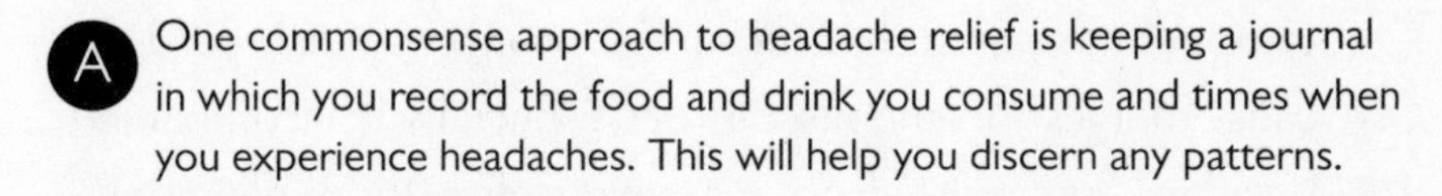

A One commonsense approach to headache relief is keeping a journal in which you record the food and drink you consume and times when you experience headaches. This will help you discern any patterns.

HOT AND SPICY SOUP

Q I read in your column that gumbo soup could help some people with migraine headaches. I started to come down with a migraine while I was on a trip and ordered some gumbo soup. The discomfort was gone within a few hours, but I don't want to eat out every time I get a headache. How do you make gumbo soup?

A There are many recipes for great gumbo, and you can select any that tickle your taste buds. Common ingredients include okra, onion, garlic, celery, green pepper, tomatoes, shrimp, and most important, chili peppers. We think the key ingredient against migraines is capsaicin (the "hot" in hot peppers). About a year ago, one man wrote to tell us about his experience with gumbo as a treatment for occasional migraines that disturb his vision. They generally occur in clusters over a period of several days: "In my town, there is a restaurant that serves very good, very spicy seafood gumbo. While waiting for my gumbo to be served one day, I noticed the onset of a migraine. My vision had deteriorated, so I could barely read just as my gumbo was served.

As I sipped my soup (it was both very hot and very spicy), my vision cleared and the headache disappeared. There was no recurrence."

We have heard from others who have also had good results with a hot and spicy soup relieving migraine symptoms. While it may not work for everyone, there is some research to support the value of capsaicin for this condition.

Heartburn

Doctors have been writing about treating heartburn for most of recorded history. In 400 B.C. the Greek physician Hippocrates noted that eating cheese after a meal could cause indigestion and discomfort, especially if accompanied by wine. Apparently Europeans were already enjoying that habit if they didn't suffer reflux. Heartburn, also known as gastroesophageal reflux disease (GERD), can have serious consequences and should not be ignored. Drugs that doctors prescribe for the condition can be very difficult to discontinue, however, and have potential side effects. We're not convinced that they are always better than home remedies.

ALMONDS

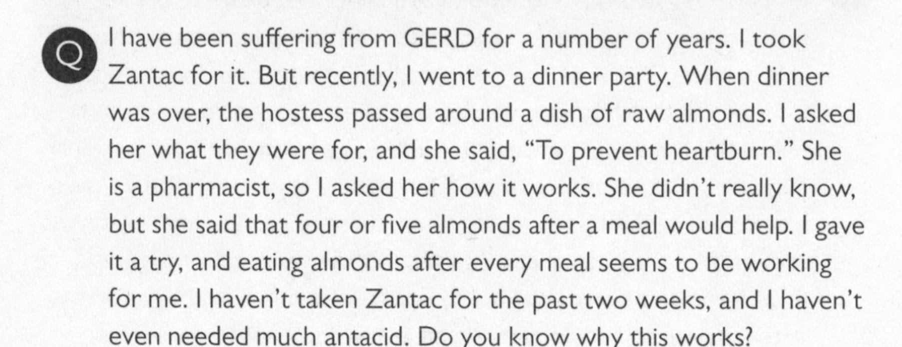

Q I have been suffering from GERD for a number of years. I took Zantac for it. But recently, I went to a dinner party. When dinner was over, the hostess passed around a dish of raw almonds. I asked her what they were for, and she said, "To prevent heartburn." She is a pharmacist, so I asked her how it works. She didn't really know, but she said that four or five almonds after a meal would help. I gave it a try, and eating almonds after every meal seems to be working for me. I haven't taken Zantac for the past two weeks, and I haven't even needed much antacid. Do you know why this works?

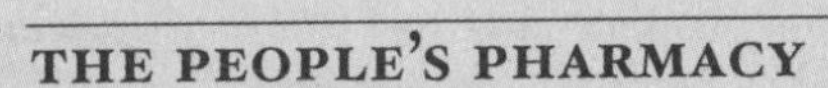

Favorite Food #12: Almonds

Most health-conscious people have heard that walnuts are a heart-healthy food. That's in part because they're rich in omega-3 fatty acids. We're all for walnuts, but we are also quite fond of almonds.

Almonds used to suffer from a bad reputation. Many people believed they were fattening and contributed to high serum cholesterol levels. But almonds have been rehabilitated. They are rich in healthful monounsaturated fatty acids and an asset in a cholesterol-lowering dietary portfolio.

Eating nuts can help lower bad LDL cholesterol and triglycerides. Recently investigators pooled data from 25 experimental studies carried out in 7 countries.[1] They found that almonds, hazelnuts, pecans, pistachios, walnuts, and even peanuts help improve blood lipids. Consuming about two and a half ounces daily helped people reduce total cholesterol by almost 11 points. Bad LDL cholesterol dropped by about 10 points, and triglycerides, a measure of fat in the blood, went down by 21 points, or nearly 10 percent.

Benefits were most apparent in people with elevated levels of triglycerides or cholesterol. The more nuts people ate, the stronger the effect. Of course, nuts have calories, so overindulging can be counterproductive by leading to weight gain. Nevertheless, it's a good idea to eat nuts instead of snacks or dessert.

Studies in Canada demonstrated that adding almonds to a vegetarian diet low in fat and rich in soluble fiber can lower total and bad LDL cholesterol as much as the prescription drug lovastatin.[2] An almond-rich diet also can modestly lower blood pressure.[3] Eating almonds can even help prevent a post-meal spike in blood sugar. One small but well-controlled study showed that eating one, two, or three ounces of almonds with a slice of white bread reduced the rise in blood sugar produced by the white bread alone.[4]

A This remedy is new to us, but it sounds safe, and almonds are one of our favorite foods. Taken in moderation, they can help lower bad cholesterol and control spikes in blood sugar. Another reader says that a bit of apple after a meal can prevent heartburn: "I have had this problem for years and recently stopped taking omeprazole so I could try to deal with this ailment in a better way. I noticed one day that food that usually gave me heartburn hadn't and realized I had eaten an apple that day. Another time after I already had heartburn symptoms, I ate three or four bites of apple and that stopped it."

BAKING SODA

Q I used to have very bad heartburn until I remembered a home remedy my mother used to make. I mix a couple ounces of water, an ounce of apple cider vinegar, and a teaspoon of sugar. After the sugar dissolves, I add half a teaspoon of baking soda, stir it briefly, and drink the mixture immediately. This offers fast relief.

A Baking soda is a time-honored approach to neutralizing stomach acid that has splashed into the esophagus and is causing heartburn.

BANANA

Q I have suffered from GERD for several years. One night, when dealing with a bad session, I ate a banana. I have no idea why; I certainly didn't expect any result. Within 30 minutes, I was able to go back to sleep. Since then, whenever a bad episode of heartburn occurs, I eat one or two bites of banana and the problem goes away. Doctors have no explanation for this. Nonetheless, it works every time, and it's not a drug. We almost always have a banana in the house.

A We're not surprised that you have found bananas helpful. Doctors in India have prescribed bananas or banana powder to treat indigestion and stomach upset from aspirin. According to a study published long ago, banana powder relieved indigestion in 75 percent of patients.[1]

BROCCOLI

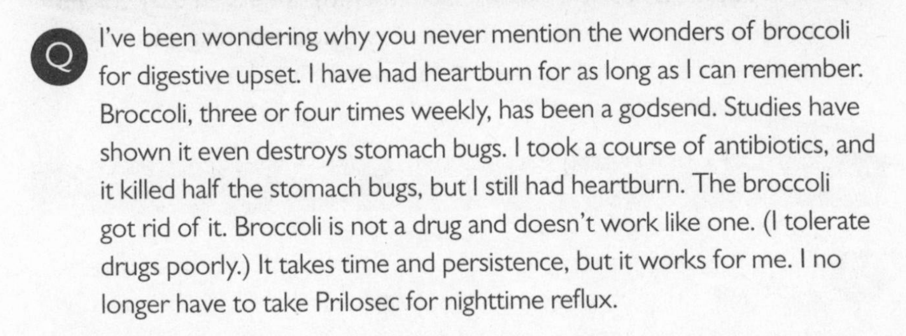

Q I've been wondering why you never mention the wonders of broccoli for digestive upset. I have had heartburn for as long as I can remember. Broccoli, three or four times weekly, has been a godsend. Studies have shown it even destroys stomach bugs. I took a course of antibiotics, and it killed half the stomach bugs, but I still had heartburn. The broccoli got rid of it. Broccoli is not a drug and doesn't work like one. (I tolerate drugs poorly.) It takes time and persistence, but it works for me. I no longer have to take Prilosec for nighttime reflux.

A Broccoli is certainly a nutritious vegetable, loaded with vitamins A, C, and K; folate; and fiber. As you note, it also contains a natural compound, sulforaphane, which can destroy *Helicobacter pylori,* a type of bacteria that live in the stomach and cause ulcers. In 2002 scientists at Johns Hopkins University reported that in test-tube studies, sulforaphane from broccoli and broccoli sprouts was able to kill *Helicobacter* inside cells, even when the bacteria had developed resistance to antibiotics. Subsequent research tested broccoli sprouts on humans. Three out of nine people who ate sprouts twice a day for a week were cured of their *Helicobacter* infections.[2]

CHEWING GUM

Q I suffer from heartburn on a daily basis, sometimes even with medication, so I was very interested to read that chewing gum might help. I take over-the-counter acid suppressors and decided to substitute sugarless gum after breakfast and then after lunch. It has been very successful for me. If I eliminate the after-dinner heartburn tablets, I run a high risk of it occurring in the middle of the night. My husband does not want to raise the head of the bed, so I take a tablet before dinner or bed to counter a nighttime episode. An additional benefit to chewing gum is that it can fight off a sweet tooth attack.

A One study published almost two decades ago documented the power of saliva to ease heartburn.[3] Sucking on a hard lozenge or chewing gum was shown to ease symptoms. Recent research demonstrates that chewing sugarless gum for 30 minutes after a meal dramatically eases acid reflux. "Gum therapy" offers an easy solution. It stimulates saliva, which buffers stomach acid, washing it out of the esophagus and back into the stomach where it belongs.

FENNEL

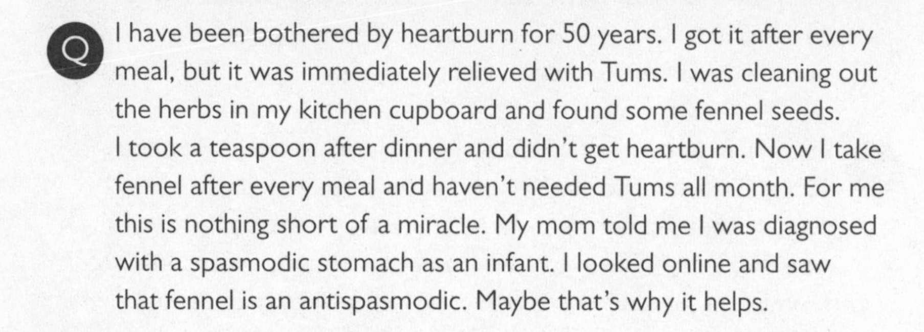

Q I have been bothered by heartburn for 50 years. I got it after every meal, but it was immediately relieved with Tums. I was cleaning out the herbs in my kitchen cupboard and found some fennel seeds. I took a teaspoon after dinner and didn't get heartburn. Now I take fennel after every meal and haven't needed Tums all month. For me this is nothing short of a miracle. My mom told me I was diagnosed with a spasmodic stomach as an infant. I looked online and saw that fennel is an antispasmodic. Maybe that's why it helps.

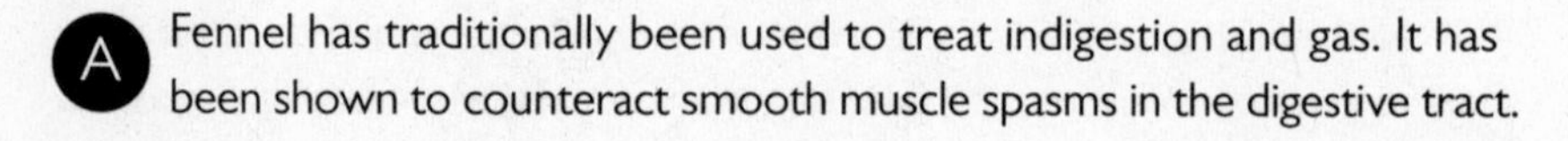

A Fennel has traditionally been used to treat indigestion and gas. It has been shown to counteract smooth muscle spasms in the digestive tract.

FAVORITE RECIPES 

Digestive Tea

Contributed by David Mathis and Debbie Mathis

½ teaspoon cumin seeds
½ teaspoon coriander seeds
½ teaspoon fennel seeds

Bring 8 to 12 ounces of water to a boil, add the seeds, and remove from heat. Steep for five minutes. Strain and enjoy. You can also mix a larger amount of seeds in equal parts and store them in a glass jar. Use about one and a half teaspoons of the mix per cup.

GINGER

Q You answered a letter from a person whose doctor prescribed Prevacid for heartburn. The patient said that she took the drug every day, even though she got heartburn only when she drank coffee. I think the physician overprescribed. Drinking arabica coffee would be better than taking Prevacid as the coffee is lower in acid. Personally, I find that hard ginger candy is very helpful in trying to deal with heartburn. You should tell people about handling heartburn without prescription drugs.

A A surprising number of foods, beverages, and medicines can aggravate heartburn. Prevention is always preferable to prescription medicines. Some people may find arabica coffee easier to tolerate, but others get heartburn from any kind of coffee, even decaf. Alcohol, fried foods, peppermint, chocolate, Valium (diazepam), and progesterone are just a few common triggers.

Q I've been suffering from a constant swollen, sore throat due to acid reflux. I've been on several different acid-suppressing drugs that worked temporarily and then stopped working. Today I tried ginger candy to soothe my throat, and it's working. Have you heard of ginger helping with reflux symptoms?

A Ginger has a long-standing reputation for soothing stomach disorders. Chinese sailors have used it for motion sickness for at least 1,000 years, and many readers have found it helpful for upset stomach. Several years ago we heard from a reader who discovered that a cinnamon-ginger drink helped her heartburn: "My reflux became really bad when I stopped hormone replacement therapy. Acid-suppressing drugs worked great, but after two months I couldn't stop them without the heartburn recurring. One night, I took colleagues to dinner at a Korean restaurant. Someone ordered persimmon punch, a concentrated cinnamon-ginger drink, for dessert. A few sips later, I felt fantastic. After a month of adding three tablespoons of the cinnamon-ginger drink to my tea morning and night, my heartburn was under control."

THE PEOPLE'S PHARMACY

Favorite Food #13: Broccoli

Although there was a time when broccoli was notoriously unwelcome on many American menus—including former President George H. W. Bush's—it is nevertheless high on the list of favorite foods of many nutrition scientists. Broccoli is one of our favorites, too, whether lightly steamed or stir-fried with a bit of garlic and a splash of soy sauce.

Broccoli is not only delicious, but also rich in nutrients: vitamins C and K, carotenoids for building vitamin A, potassium, folic acid, and fiber. We also have heard from several readers that a diet featuring broccoli can help reduce heartburn. Research indicates that benefit might be due to sulforaphane, a compound that can get rid of intestinal bacteria such as *Helicobacter pylori*.[1] Infections with this beastie are responsible for many stomach ulcers.

Research shows that sulforaphane is also active against a range of other pathogens.[2] The sulforaphane in broccoli (along with that in cabbage, brussels sprouts, and other cruciferous vegetables) may also be beneficial in lowering the potential for developing cancer of the bowel. Research on genetically engineered mice shows that sulforaphane in the diet reduced the number and size of colon polyps that the mice developed.[3]

Researchers working to find out exactly how broccoli-derived sulforaphane accomplishes this have discovered that there may be more than one pathway.[4] In addition, some basic research suggests that a tasty pairing of broccoli with garlic can provide added protection.[5]

A couple of cautions: People with hypothyroidism should be moderate in their consumption of cabbage-family vegetables, including broccoli, and should also apply the same caution in eating soybeans. Some of the common compounds in these plants can interfere with the body's utilization of thyroid hormone.

The other concern is gas. People vary in their susceptibility, but those who are especially sensitive to the gas-producing potential of broccoli should go easy. Fennel tea or Angostura bitters may come in handy for those who do overindulge.

Persimmon Punch

2 quarts water
½ cup thinly sliced fresh ginger
3 cinnamon sticks
½ cup honey
1 ripe persimmon, washed and sliced thinly

Combine ginger and cinnamon in water and simmer for 30 minutes. Strain the liquid and stir in honey and persimmon. If fresh persimmon is unavailable, one-half cup sliced dried persimmon may be substituted. Chill in the refrigerator overnight and serve cold. The punch can be stored for up to a week.

Ginger Pickle

Contributed by David Mathis and Debbie Mathis

Peeled fresh ginger root
Fresh lime juice
Salt

Slice thin cross sections of ginger. Squeeze a little fresh lime juice over the slices, and lightly sprinkle them with salt. To kick-start good digestion, eat about two slices per person, per meal, about 15 minutes before the meal.

HOT PEPPERS

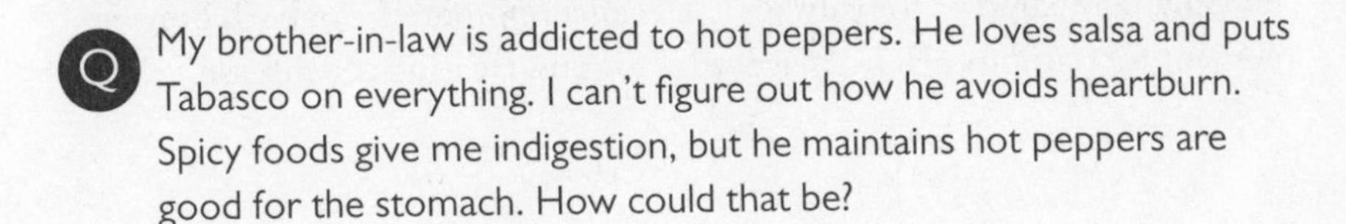

Q My brother-in-law is addicted to hot peppers. He loves salsa and puts Tabasco on everything. I can't figure out how he avoids heartburn. Spicy foods give me indigestion, but he maintains hot peppers are good for the stomach. How could that be?

THE PEOPLE'S PHARMACY

Favorite Food #14: Hot Peppers

Some people shun spicy foods. Even a hint of heat has them sweating and gasping. Then there are the pepper heads. The hotter the food, the happier they are. Faced with chicken curry, chili, eggs, gumbo, or spaghetti without a bottle of hot sauce handy, they get anxious. For them, such foods demand an extra kick.

Many believe spicy foods are hard on the digestive tract. For years physicians warned people with heartburn (also called dyspepsia) or ulcers to avoid spicy foods and to stay away from citrus fruit, alcohol, fatty foods, chocolate, coffee, and tea.

Yet in the medical literature the evidence that spicy foods lead to heartburn is thin.[1] And we are here to tell you that spice can be nice.

In a double-blind, placebo-controlled trial, researchers in Italy gave hot peppers to patients with heartburn. Instead of making them worse, the hot peppers reduced symptoms of indigestion by 60 percent, compared to roughly 30 percent for patients given a placebo. The researchers concluded, "Although larger trials with standardized materials are needed, capsaicin [the active compound in hot peppers] could represent a potential therapy for functional dyspepsia."[2]

In one study, physiologists found that capsaicin might even be beneficial for the stomach lining.[3] Other studies on rats and cats have shown that capsaicin can actually inhibit stomach acid secretion.[4] Another study demonstrated the power of the hot stuff in hot peppers to protect the stomach against aspirin-induced damage.[5]

Further research has shown that capsaicin protects against stomach damage caused by the arthritis drug indomethacin and by alcohol.[6] There is also increasing evidence to suggest that capsaicin may help prevent platelets from sticking together and causing clots.[7]

Even more exciting is the surprising amount of new research suggesting that hot peppers and capsaicin may have significant anti-cancer potential—in both preventing and treating the disease.[8]

A Your brother-in-law actually has some science on his side. Italian researchers reported that red pepper powder in capsules reduced stomachache, fullness, and nausea by 60 percent.[4] In comparison, a look-alike placebo reduced these symptoms by half as much. Scientists think the "magic" ingredient in hot peppers is capsaicin, which is used in arthritis remedies and other creams to relieve pain. Studies on rats have shown that capsaicin reduces stomach damage caused by aspirin or alcohol.

LOW-CARB DIET

Q For many years I suffered from heartburn and took Alka-Seltzer four or five days a week at bedtime. I had been advised to avoid fatty foods, so I ate a very low-fat diet with lots of rice, pasta, and beans. I ate nothing fried. Then, last year, I made a New Year's resolution to lose weight, and I tried the Atkins diet. I worried that eating greasy food like sausage, eggs, and cheese would aggravate my heartburn, but I decided to try the diet anyway. (I could no longer button my jeans.) I lost 25 pounds in 10 weeks and have kept the weight off for 15 months. I also lost my heartburn, even before the weight came off. I no longer take any antacid. Have others reported this benefit?

A We have heard from some people that the Atkins diet helps relieve symptoms of heartburn and acid reflux. There is even a preliminary medical report documenting five cases of patients whose acid reflux disappeared when they adopted a carbohydrate-restricted diet.[5] Research also suggests that a very-low-carbohydrate diet may help ease some symptoms of irritable bowel syndrome.[6]

Q I once suffered from GERD and took Prilosec daily. Even so, I had severe heartburn, and food often got lodged in my esophagus, even after it was surgically stretched. I had a few other minor health problems and was a bit overweight, so I was ready to make some changes. A friend suggested a low-carb diet. Though I expected little, I opted to try it. After three days, I realized that I had no heartburn, so I discontinued the Prilosec without ill effects. I'd tried to stop the drug before, but the heartburn had gotten worse. A short time later,

I found I could swallow without choking. After years of suffering, my life was normal once again. It has now been a year, and I have lost 45 pounds. I'm still fine. My message is simple: If you are having heartburn, GERD, or swallowing difficulties, consult your doctor to find out about a temporary low-carb diet. I strongly believe it will help many people.

A Thanks for sharing your story. In one study, people on a carbohydrate-restricted diet had significantly less heartburn.[7]

PAPAYA

Q I frequently have heartburn and finally found a wonderful remedy: papaya pills. Every time I have heartburn, I eat one of the pills and the heartburn disappears. My doctor says it's fine to use them. Others might like to know about this great way to treat heartburn.

A Papaya is a tropical fruit that contains an enzyme (papain) that may be very helpful for digestion. Although papain does nothing to suppress acid, some people report that papaya relieves heartburn. Anyone who is allergic to latex should avoid papaya since there is cross-reactivity between latex and papaya, which could be very dangerous. Papain may also increase the blood-thinning effects of warfarin (Coumadin).

VINEGAR

Q I have been on acid-blocking drugs for years to treat heartburn. Initially I was on Prilosec, then Prevacid, and now Nexium. I recently saw a naturopathic physician who said that stomach acid is necessary for proper digestion and good health. He said I could be more vulnerable to infections if I keep shutting down my stomach acid and recommended two tablespoons of apple cider vinegar in a glass of water instead. This seems totally illogical to me.

A Vinegar seems an odd remedy for heartburn, but this isn't the first time we have heard of it. One reader reported that his doctor recommended a tablespoon or two of vinegar in water for heartburn relief. He tried it, and it worked. Studies have suggested that constantly suppressing stomach acid may increase the risk of pneumonia or severe infectious diarrhea.[8] Acid in the stomach kills bacteria. Without it germs may survive and cause trouble. Nonetheless, some people require such medication to avoid scarring the esophagus.

Q I'm trying to find out what causes heartburn and how best to treat it. Is it caused by eating too fast? Can heartburn cause heart attacks? My boyfriend has discomfort nearly every night. A guy he works with has recommended vinegar, but that seems ridiculous.

A Heartburn happens when stomach acid splashes back into the esophagus. The corrosive chemicals are irritating to the delicate lining of the gullet. Though heartburn may sometimes feel like a heart attack, it does not cause one. If there is any question about the cause of chest pain, it warrants a visit to the ER for checking out. Many foods and drugs can make heartburn worse. Eating rapidly or overeating may also aggravate it. We agree that vinegar doesn't sound like a logical thing to take for heartburn, but we have heard from several readers who found it helpful.

YELLOW MUSTARD

Q My wife and I both use plain old yellow mustard to combat indigestion or acid reflux. It works very well for us. If we swallow a spoonful of mustard before an Italian meal, we are okay.

A Although mustard may seem like the last thing anyone would want to take for heartburn, we have heard from others that it can be helpful. The turmeric that makes mustard yellow was traditionally used for digestive upset in Chinese medicine. Mustard also contains vinegar, which some people find helpful against heartburn.

Hemorrhoids

Hemorrhoids afflict lots of people, especially older adults and pregnant women. And as anyone who's ever suffered from them knows, they can be very painful. We have now heard from several readers who have told us that eating a teaspoon or two of blackstrap molasses daily helps alleviate their suffering. We have no idea why this might be, and we haven't been able to find any scientific research to support this unconventional use for molasses. But we are delighted to hear that it has helped some folks get relief.

BLACKSTRAP MOLASSES

Q Recently a reader wrote you touting blackstrap molasses as a treatment or cure for hemorrhoids. I am a hemorrhoid sufferer. I've used over-the-counter medications and the prescription ointment Proctosol-HC without much success. My problem seems to flare up when I eat certain foods, especially spicy dishes, and when I engage in vigorous exercise such as running or bike riding. I decided to try taking blackstrap molasses, one teaspoon twice a day. Remarkably, in less than a month I have found relief! The condition has improved at least 75 percent, with only occasional minor irritation. The burning, itching, and even blood spotting have all but disappeared. Believe me, I was skeptical about this remedy, so even after some initial relief, I purposely ate foods that normally cause a flare-up. Much to my surprise, there was no outbreak. Has anyone else reported success?

A **Your testimonial is the first we have received. We too would be interested to learn of others' experience.**

Q I read your column about blackstrap molasses for hemorrhoids about three weeks ago. I too had huge, painful, bleeding hemorrhoids that weren't responding to Proctosol-HC, sitz baths, or ice. Even though I was somewhat skeptical, I bought some blackstrap molasses and started taking a teaspoon twice a day. Lo and behold, within a week the hemorrhoids and pain diminished, and within two weeks my hemorrhoids had all but disappeared. I am amazed and thrilled. I continue to take the blackstrap molasses every day, but I have reduced the amount to one teaspoon a day. I will stop taking the molasses, but I am not sure when. Thanks so much for writing about it. I never would have known about this treatment and would have ended up with surgery. A miraculous cure, and the timing was phenomenal.

We continue to be astonished that people find blackstrap molasses eases hemorrhoids. It might be coincidence, but if it works, it's a low-cost, low-risk remedy. Those who must avoid sugar should be cautious since molasses is a by-product of sugar manufacturing.

Hiccups

Hiccups are almost always treated with home remedies. Doctors would be hard-pressed to find drugs they would want to prescribe for singultus, the medical term for hiccups. Almost everyone knows about drinking from the wrong side of the glass (you have to bend over) or swallowing a spoonful of sugar. Listeners and readers have also come up with some more unusual hiccup remedies.

CHOCOLATE CHIPS

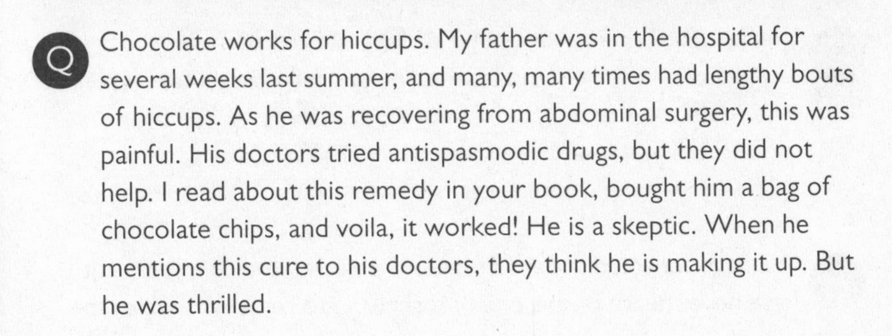

Q Chocolate works for hiccups. My father was in the hospital for several weeks last summer, and many, many times had lengthy bouts of hiccups. As he was recovering from abdominal surgery, this was painful. His doctors tried antispasmodic drugs, but they did not help. I read about this remedy in your book, bought him a bag of chocolate chips, and voila, it worked! He is a skeptic. When he mentions this cure to his doctors, they think he is making it up. But he was thrilled.

A We first heard about eating a few chocolate chips to stop hiccups from a radio show listener. Her Danish grandmother had always used this remedy, and she continued the tradition with her children.

DILL PICKLE JUICE

Q The best hiccup remedy I have found is dill pickle juice. One to two ounces does the trick.

A You are not the first reader to sing the praises of drinking pickle juice as a way of getting rid of the hiccups. Perhaps the salt or the vinegar is responsible. Pickle juice is high in sodium, however, so people with high blood pressure or heart failure should stay away from this home remedy.

DRINKING THROUGH PAPER

Q When I was in college a fellow student gave me this silly remedy for hiccups. Place a paper napkin over the top of a glass of water. Have the hiccuper take sips of water through the napkin. It works!

A Before the days of paper napkins, people used a clean cloth handkerchief in one variant of this hiccup remedy. We have no idea why this would work, but readers assure us that it does.

EAR PULLING

Q I have a hiccup cure that I've used all my life and passed on to my kids: Have someone stand behind you and pull straight up on your ears while you take sips of water. I do this by myself by taking a mouthful of water and swallowing it in small amounts while pulling up on my ears. I kid you not, it works every time!

A We have collected dozens of hiccup remedies over the years but have never heard of this one. It is similar to a remedy in which the hiccuper drinks water while someone stands behind him and pushes on the little flaps at the front of the ears.

Q Here is a cure for hiccups. The person with hiccups plugs up both ears with her fingers and drinks water. This is easiest with the help of another person holding the ears, but by using a bottle of water you can accomplish this by yourself.

A It helps to have someone press on the flap, called the tragus, that covers the opening of each ear canal.

GREEN OLIVES

Q I have used one or two green olives for hiccups for many years, as have all my family members. I don't know why it works, it just does. It stops the hiccups almost immediately. This remedy even worked on a friend who had been through surgery and had suffered for three weeks until I gave him an olive. Have you ever run across anyone else who has used this? I'm rather curious as to why it works.

A We have been collecting hiccup remedies for more than 30 years, but this is the first time we have heard of using green olives. Most hiccup remedies work by stimulating the phrenic nerve at the top of the hard palate. Chewing crushed ice or swallowing a spoonful of granulated sugar or ice cream seems to interrupt the hiccup reflex. Perhaps the green olives work on a similar principle.

Q I am writing about your article on hiccups. Recently when I had a bad case of the hiccups, I remembered reading about eating a couple green olives. I have to tell you I am a firm believer now. The hiccups were gone almost instantly after I ate the second olive. What a wonderful idea!

A We appreciate your testimonial. We don't know if it is the olive itself or the vinegar in the brine that does the trick by stimulating the phrenic nerve, but we are always pleased to learn that a hiccup remedy has helped.

Q I just wanted you to know I recently read your column about the power of green olives to fight hiccups. My five-year-old got the hiccups the next day. Guess what? Just one green olive did the trick. Coincidence or science? Who knows, but we're quite convinced that it worked!

A We're delighted to learn that this unusual remedy worked for you.

PEANUTS

Q My favorite hiccup cure, peanut butter, has come under fire as a choking hazard. I found a substitute that works just as well, but without the risk: peanuts! If you're allergic to peanuts I wouldn't suggest this, but if you've used the peanut butter hiccup cure before, this works too. I gave up peanut butter as part of my weight loss plan, so I don't keep any in the house. I do sometimes like to snack on dry-roasted, unsalted peanuts. I got the hiccups one night, and I ate a handful of peanuts and drank a little water. My hiccups were gone! A friend had the same result, so I expect this will help others.

A Swallowing a spoonful of dry sugar and sucking on a section of lemon are time-honored hiccup cures. Thanks for a new one.

PINEAPPLE JUICE

Q I am surprised that I have never seen pineapple juice mentioned in your column as a cure for hiccups. It works with one sip!

A Thanks for mentioning it. We've heard that pineapple juice can ease joint pain, but you are the first to report that it cures hiccups.

VINEGAR

Q I have been using vinegar as a remedy for hiccups for over 20 years. I know of no instance when a teaspoon of vinegar did not eliminate even the most stubborn case. Usually I use white vinegar, but I have used balsamic and rice vinegar with success. I believe this answers the question of why green olives cure hiccups. It is not the olive but the vinegar in the brine!

Thanks for your remedy and the answer to a puzzle.

High Blood Pressure

Elevated blood pressure increases the risk for two major killers—stroke and heart attack. High blood pressure also plays a role in dementia and kidney damage. That's why doctors emphasize keeping blood pressure under control. Blood pressure varies throughout the day, so it makes sense to measure it at home from time to time rather than relying on a single measurement or two at the doctor's office. Exercise raises blood pressure, but afterward pressure generally drops noticeably. To get a good reading at home, make sure your arm is resting at about the same level as your heart. Use a cuff that is the correct size for your arm. Don't talk while you are taking your blood pressure, and write the number down in a log. Keeping records will help you and your doctor figure out how best to handle your blood pressure.

BEET JUICE

Q Is there anything in the way of vitamins or herbs that a person can take instead of a prescription drug for high blood pressure? I've heard about garlic, but I don't like it much. Is there anything else?

A The newest candidate for natural blood pressure control is beet juice. Beets are high in dietary nitrate and increase nitric oxide in the body. Nitric oxide helps blood vessels relax (thus helping to

THE PEOPLE'S PHARMACY

Favorite Food #15: Beets

Distinctly colorful—the expression "beet red" comes to mind—this old-fashioned root vegetable was adopted by eastern Europeans long ago. Red beets (they also come in gold) make a delicious addition to salads, and there's nothing like borscht, served hot in winter or cold with a dollop of sour cream in summer. (And don't worry if your urine or stool turns red; it is simply a natural consequence of eating beets.)

Beets made headlines back in 2008, when British investigators announced the findings of a study on beet juice and blood pressure. They had randomly assigned volunteers to drink either two cups of beet juice or two cups of water. Blood pressure was carefully monitored before and up to 24 hours after the subjects drank the beet juice.

In the British study, beet juice lowered blood pressure readings by around ten points, and the effect lasted for nearly a day.[1] This benefit compares well to many antihypertensive drugs.

This is good news for people who take blood pressure medicine, since a recent article has raised questions about the safety of several of them. The article noted that Atacand, Diovan, and Micardis may be linked to "a modestly increased risk of new cancer occurrence."[2] Another recent study suggests that about 8.5 ounces of beet juice can significantly lower systolic blood pressure.[3]

Scientists have hypothesized that the nitrate content of beets is responsible for this effect. (Spinach and other dark, leafy greens are also naturally rich in nitrate.) Nitrate in the diet leads to nitric oxide formation in the tissues. That, in turn, helps blood vessels relax and discourages blood clot formation.

Research on mice suggests that nitrate-containing foods can reduce the inflammation triggered by high blood levels of cholesterol.[4] And, in fact, some of our readers have indeed reported successfully lowering their LDL cholesterol by adding beets to their diet.

lower blood pressure), has anti-inflammatory properties, and discourages blood clot formation. A diet rich in vegetables and dark chocolate can also lower blood pressure. Pomegranate and grape juice, magnesium supplements, and breathing exercises also can be beneficial.

CHOCOLATE

Q I started eating Hershey's dark chocolate when it was on sale a few weeks ago. I enjoy about five of the little squares twice a day. My systolic and diastolic blood pressure numbers went down about 15 or 20 points each.

A Chocolate will never be a substitute for blood pressure medicine, but some data support your experience. Studies have demonstrated modest benefits of cocoa and dark chocolate in lowering blood pressure.[1] Your reaction to chocolate is greater than average. The amount used in studies ranges from a little less than 10 grams (one small Ghirardelli square) to 100 grams (a Ritter Sport bar). Keep in mind that chocolate is high in calories, and weight gain can drive blood pressure up.

DASH DIET

Q Many years ago, I read that people should eat brown rice to reduce blood pressure. More recently I read that the rice diet works, but modern medicines are better for treating high blood pressure. Have you ever heard of this diet regimen?

A The rice diet (mostly brown rice, vegetables, and fruit; no salt) was originally developed to control blood pressure, before many medicines were available. Since then, people have also used it successfully for weight loss. Research shows that a diet rich in vegetables and very low in sodium (the DASH diet) can help control blood pressure almost as well as some medications. DASH stands for Dietary Approaches to Stop Hypertension. In one study, women who followed this diet were

THE PEOPLE'S PHARMACY

Favorite Food #16: Chocolate

One of our favorite foods is chocolate, the darker the better. A surprising amount of research shows that the active ingredients in chocolate and cocoa have exciting biological benefits. Cocoa flavonoids, as these compounds are called, reduce insulin resistance and improve insulin sensitivity.[1] These are important issues for people with type 2 diabetes, because their bodies often make normal or even high amounts of insulin, but their cells become resistant to this hormone. Increasing insulin sensitivity can help improve blood sugar control.

Eating dark chocolate may also reduce inflammation, now being blamed for everything from diabetes and heart disease to cancer and Alzheimer's. One marker of inflammation is C-reactive protein (CRP). Scientists have found that eating less than an ounce of dark chocolate every three days significantly lowered CRP levels.[2] Eating a lot more than this amount was counterproductive.

Epidemiologists in China analyzed eight studies to see if cocoa polyphenols in dark chocolate can lower cholesterol.[3] They found that these compounds lower bad LDL cholesterol and total cholesterol by approximately six points. But the benefit was seen only in people at risk of heart disease. Research suggests that cocoa flavonoids can also lower blood pressure, improve flexibility of blood vessels, and keep blood platelets from sticking together to form blood clots.[4] They may even reduce the chance of a heart attack, which is, after all, why we are concerned about blood clots or blood pressure. A Swedish study followed more than a thousand heart attack survivors for eight years. A person who has had one heart attack is at increased risk for a second one. The investigators found a striking result: Compared to people who never ate chocolate, those who indulged twice a week or more were about three times less likely to have a second heart attack.[5] If a pricey prescription drug produced such dramatic results, physicians would be prescribing it like candy!

less prone to heart disease.[2] Other natural approaches for controlling blood pressure include weight loss, stress management, deep breathing, tea, Concord grape juice, pomegranate juice, dark chocolate, and minerals such as potassium and magnesium

EXERCISE

Q I am a 63-year-old female. A few months ago, I finally yielded to my doctor's pressure and went on a blood pressure medication, against my better judgment. The medication is metoprolol succinate. If you could tell me some of its side effects, I would be grateful. I am not feeling well, and I am guessing this new med is the culprit.

A Metoprolol, like other beta-blocking blood pressure medicines, may cause fatigue, dizziness, and diarrhea. Some people develop an itchy rash, while others may find themselves short of breath. It slows heart rate, sometimes by quite a lot. Many cardiologists are reassessing beta-blockers. Current studies show that such drugs are rarely considered first line treatments for hypertension.[3]

Blood pressure control is very important, so don't stop your medication on your own. (It is dangerous to stop a beta-blocker suddenly.) Discuss your symptoms with your doctor and ask about other possible treatments. Perhaps this reader's experience will help you: "I was told I need to be on blood pressure meds. I don't like to take pills, so I tried breathing exercises and going to the gym. I didn't have high blood pressure at my last visit."

GARLIC

Q An old doctor told my husband that Kyolic garlic capsules might lower his blood pressure. I started taking them myself, morning and night. My blood pressure is now lower than normal most of the time. My doctor has taken me off the atenolol and lisinopril I used to take.

A You did this experiment properly, with a doctor's supervision. People should not stop blood pressure medication on their own. Studies on garlic and blood pressure have had mixed results. A recent review and meta-analysis from Australia concludes that garlic works better than placebo in lowering blood pressure.[4] A recent Russian study found that a timed-release garlic formulation worked better than immediate-release standardized garlic capsules.[5]

GRAPESEED EXTRACT

Q I've been hearing about grapeseed extract as an antioxidant. What do you know about it?

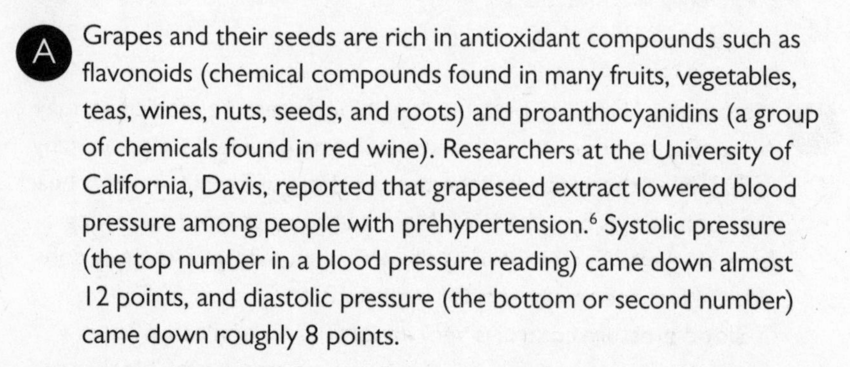

A Grapes and their seeds are rich in antioxidant compounds such as flavonoids (chemical compounds found in many fruits, vegetables, teas, wines, nuts, seeds, and roots) and proanthocyanidins (a group of chemicals found in red wine). Researchers at the University of California, Davis, reported that grapeseed extract lowered blood pressure among people with prehypertension.[6] Systolic pressure (the top number in a blood pressure reading) came down almost 12 points, and diastolic pressure (the bottom or second number) came down roughly 8 points.

GREEN AND OOLONG TEA

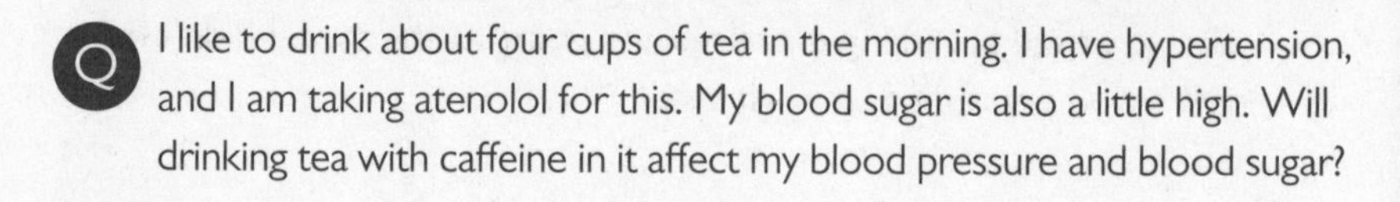

Q I like to drink about four cups of tea in the morning. I have hypertension, and I am taking atenolol for this. My blood sugar is also a little high. Will drinking tea with caffeine in it affect my blood pressure and blood sugar?

A A cup of black tea has about 40 or 50 milligrams of caffeine, depending on how long it steeps. So in your four cups you are getting roughly 160 to 200 milligrams of caffeine, roughly as much as you'd get in two cups of coffee. Research suggests that consuming green or oolong tea (which contain less caffeine than black tea) may help prevent high blood pressure.[7] In general, however, caffeine can raise blood pressure somewhat, especially for people who are also under stress. Another study showed that caffeine on an empty stomach does not have an impact on blood sugar, but if taken with a meal, caffeine can raise blood sugar and insulin levels in type 2 diabetics.[8] Maybe you should make sure you drink your tea mid-morning without a snack.

SALT

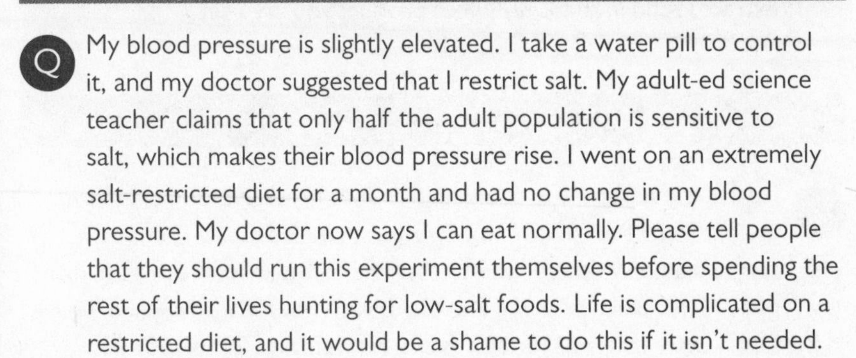

Q My blood pressure is slightly elevated. I take a water pill to control it, and my doctor suggested that I restrict salt. My adult-ed science teacher claims that only half the adult population is sensitive to salt, which makes their blood pressure rise. I went on an extremely salt-restricted diet for a month and had no change in my blood pressure. My doctor now says I can eat normally. Please tell people that they should run this experiment themselves before spending the rest of their lives hunting for low-salt foods. Life is complicated on a restricted diet, and it would be a shame to do this if it isn't needed.

A The link between salt (sodium chloride) and high blood pressure has been controversial for decades. Studies suggest that cutting back on salt can lower blood pressure modestly. A carefully run, long-term trial indicated that substantial sodium reduction lowers blood pressure only two or three points.[9] Some people are especially salt sensitive and benefit from a sodium-restricted diet. Others, like you, don't see any improvement. In a comprehensive study, sodium restriction led to a 25 percent decline in heart attacks and strokes.[10] Though cutting back on salt may not make a big difference for everyone, it clearly can help some people control cardiovascular disease.

STEVIA SWEETENER

Q In your book *Best Choices from the People's Pharmacy*, you said that the no-calorie sweetener *Stevia rebaudiana* was not approved for use in the United States. I have some great news for you! I found it in a nonsugar sweetener in the supermarket under the brand name of OnlySweet. It is made by Sunwin, and their website is *www.onlysweet.com.* I am a type 2 diabetic with high blood pressure. This sweetener not only lowers my blood sugar, but also helps reduce my blood pressure. I am enthusiastic!

A Thanks for the alert. The FDA has approved a stevia extract, but the product that the FDA approved, rebaudioside A, does not seem to lower blood sugar or blood pressure in healthy adults.[11] A different stevia extract, stevioside, brought both blood pressure and blood sugar under control in rats. We were unable to determine the extract found in the OnlySweet product you are using.

TURMERIC

Q I learned that turmeric could help reduce my high blood pressure. It went from 160/80 to 140/60 after an Indian meal. Then my nutritionist daughter advised me to use one-half teaspoon turmeric daily. I add it to a clove of garlic and blend with fruits in a soy shake. I have lost weight, and my blood pressure runs about 109 to 120/65.

A What an interesting discovery. We don't think doctors will prescribe curry, but research on rodents suggests that curcumin, the active ingredient in turmeric, can lower blood pressure.[12] Turmeric is the yellow spice used in curry and yellow mustard. Although yours is the first report on using this spice for hypertension, readers have found a number of other uses for it. We have heard stories about benefits for arthritis, bursitis, diabetes, gout, leg cramps, and psoriasis. Some may be allergic to turmeric. Further, this herb may interact with the blood thinner warfarin (Coumadin) and cause dangerous bleeding.

High Cholesterol and Triglycerides

High cholesterol is a risk factor for heart disease, though not the only important one. High triglycerides—blood fats that function somewhat independently of cholesterol—are also harbingers of heart disease. Several medications lower cholesterol effectively. Some even reduce the risk of heart attack or stroke. But many cholesterol-lowering drugs have side effects. Statin-type cholesterol-lowering medicines like atorvastatin (Lipitor) or simvastatin (Zocor) may cause muscle pain, weakness, or even a life-threatening muscle breakdown called rhabdomyolysis, which also threatens the kidneys. Drugs that are not statins, such as Zetia (included with a statin in Vytorin), may also cause muscle pain.

Few cholesterol-lowering drugs reduce triglycerides. Consequently, many people are eager to learn of home remedies to try for lowering high triglycerides and cholesterol levels.

CERTO AND GRAPE JUICE

Q After a year, I got my cholesterol down from over 260 to below 220. Then sore joints led me to glucosamine and chondroitin. Now my cholesterol is over 240. I think there is a connection. My doctor recommended Certo and grape juice to lower cholesterol and possibly ease arthritis pain. How much do I need, and is there anything else natural I could try?

A Research has not confirmed that glucosamine and chondroitin raise cholesterol, but other readers have reported a similar problem.

You may be interested to note that grape juice has been shown to lower cholesterol. Soluble fiber in the form of liquid plant pectin (Certo) may have a similar benefit. Many readers tell us the combination also helps ease arthritis pain. One recipe involves adding two teaspoons of Certo to three ounces of grape juice and drinking it three times daily. Fish oil also can help lower triglycerides and ease joint pain, while psyllium can reduce cholesterol levels.

CINNAMON

Q Taking lovastatin has controlled my total cholesterol and bad (LDL) cholesterol. But my triglycerides were always very high. My doctor had no suggestions, so I decided to try cinnamon. After I started taking cinnamon, my triglycerides went from 350 all the way down to 150 in four months. I took one-fourth to one-half teaspoon daily with a glass of water.

A Thanks for sharing your extraordinary results. We have heard from other readers who have managed to lower cholesterol and blood sugar with a daily dose of cinnamon. Some people report that this spice causes heartburn. We would encourage anyone who considers cinnamon to treat it as a drug and to check with a physician for monitoring of liver enzymes.

FISH OIL

Q I was saddened to read about people suffering from the side effects of cholesterol-lowering statin drugs. In less than a year, I have lowered my cholesterol 30 points by doing nothing other than taking a tablespoon of cod-liver oil a day. It not only lowers bad cholesterol but also raises good (HDL) cholesterol. I wish more people knew about this very simple and safe remedy. One friend said it tastes bad, but that's not true if you take Carlson or Nordic Naturals.

A The benefits of fish oil are well established. There are more than 10,000 articles on fish oil in the medical literature. Many refer to its abilities to lower triglycerides and to raise HDL cholesterol. There is even a purified fish oil available as the prescription drug Lovaza. Fish oil also has anti-inflammatory properties.

OATMEAL AND BARLEY

Q Oatmeal and barley have lowered my cholesterol significantly. How do they work?

A Oatmeal and barley are rich sources of soluble fiber. They bind to cholesterol in the digestive tract and keep it from being absorbed.

THE PEOPLE'S PHARMACY

Favorite Food #17: Fish and Fish Oil

Fish is widely considered one of the most valuable foods you can eat. In just one week in 2002, three of the country's most prestigious medical journals published separate studies demonstrating the same conclusion: People who eat more fish or take fish oil are less likely to die of a heart attack.[1] Fish oil may help stabilize the electrical activity of heart cells, thereby lowering the risk of potentially life-threatening heart rhythm disturbances. It may work better as a preventive. A recent study showed daily consumption of margarine containing almost 500 milligrams of elcosapentaenoic acid, or EPA, and docosahexaneoi acid, or DHA—the fatty acids in fish oil—did not prevent second heart attacks.[2] According to the American Heart Association, two servings of oily fish per week provide roughly 400 to 500 milligrams of omega-3 fatty acids per day.

Research also suggests that including fish in the diet seems to protect the brain from strokes.[3] So the old idea that fish is brain food appears to be true. A study of nearly 15,000 elders in China, Cuba, Dominican Republic, India, Mexico, Peru, and Venezuela found populations with higher fish intake had lower rates of dementia. The authors conclude that the epidemiological findings in many low- and middle-income countries are consistent with the "neuroprotective action" of omega-3 fatty acids.[4]

People who eat more fish are less susceptible to age-related macular degeneration, a leading cause of blindness. Collecting dietary data from some 2,500 people for years in the Blue Mountains Eye Study, Australian investigators found one serving of fish each week reduced the risk of developing early age-related macular degeneration by about 31 percent.[5]

Our readers also tout fish oil for easing morning stiffness, arthritis, and aches and pains. One reader wrote, "My husband is under a doctor's care for arthritis. Within a day of starting fish oil, swelling began to go down in his fingers. He still takes prescribed medication, but in lower doses. This relief is so much better."

PLANT STANOL ESTERS

Q I was surprised that you didn't include a reference to plant stanol esters in your response to a question about lowering cholesterol. Six years ago my husband's cholesterol was at 385. He didn't want to take statins because of interactions with other health problems. His cardiologist recommended that he use at least one tablespoon of Benecol at each meal and make some other dietary changes. This brought his cholesterol down to 185, and the balance of HDL versus LDL was restored to a healthier level.

A The FDA has concluded that stanol esters can lower cholesterol and reduce the risk of heart disease. Spreads such as Benecol, Promise activ, and Take Control contain these plant products. We are delighted to learn that your husband's dietary changes had such a profound impact on his cholesterol levels.

RED YEAST RICE

Q I heard on the news that red yeast rice can help lower cholesterol, but I'd like to hear some pros and cons. Are there dangerous side effects? Someone told me red yeast rice can cause liver problems. Is that true?

A For centuries, people have used red yeast rice in food as a preservative, flavoring, and coloring agent. It makes Peking duck red, for example, and is also used in red rice vinegar. It has also been used in traditional Chinese medicine for indigestion and to promote blood circulation. Red yeast rice can also lower cholesterol. It contains compounds related to statin-type cholesterol-lowering drugs such as lovastatin (Mevacor). A recent study in China shows that red yeast rice can nearly halve the risk of a second heart attack.[1] One of our readers reported success with red yeast rice: "For years I balked at taking a statin, but my cholesterol rose in spite of attempts to control it with diet. I tried Pravachol but did not like the way I felt. My cardiologist suggested red yeast rice. Within six weeks my LDL level had dropped from 187 to 123."

THE PEOPLE'S PHARMACY

Favorite Food #18: Grape Juice

The buzz about the benefits of red wine leads many to assume that grape juice is just a wimpy alternative for teetotalers. Nothing could be further from the truth. Research suggests that purple grape juice has some pretty powerful properties.

A review of the literature by investigators at Boston University reveals that compounds in grapes and grape juice can lower blood pressure, reduce risk of blood clots by inhibiting the clumping of blood platelets, block oxidation of bad LDL cholesterol, improve flexibility of blood vessels, and reduce inflammation.[1] In a double-blind, placebo-controlled trial of Korean men with high blood pressure, real Concord grape juice lowered blood pressure significantly more than fake juice.[2] Evidence also suggests that grapes and grape juice can enhance immune function.[3]

One of the protective constituents in grapes is resveratrol. Plants make resveratrol in part to ward off bacterial and fungal infection, and this compound is especially abundant in the skin of red grapes. Research on animals suggests that resveratrol may have anticancer effects and may reduce the formation of plaque within the brain that may lead to Alzheimer's disease.

A leading nutritional expert, neuroscientist James Joseph, recently published a summary of his research in which he concluded that "a greater intake of high-antioxidant foods such as berries, Concord grapes, and walnuts may increase 'health span' and enhance cognitive and motor function in aging."[4]

Grapes also seem to work for coughs and arthritis. Over the years we have heard from hundreds of readers who sing the praises of this grape juice "cocktail" for arthritis: Combine one part apple cider vinegar with three parts apple juice and five parts grape juice, and drink four ounces daily. Alternatively, you can add two teaspoons of Certo (liquid pectin used by home canners to make jams and jellies thicken) to three ounces of purple grape juice, and drink this arthritis-fighting concoction three times daily.

Some people experience side effects, including muscle pain and weakness and liver damage, so anyone taking red yeast rice should be under medical care and have regular liver enzyme tests.

SMOOTHIE

Q My 15-year-old son had low HDL and a poor cholesterol profile at his checkup. (Total cholesterol was 146—HDL 29 and LDL 96). For ten months I have had him drink my version of the cholesterol-lowering smoothie I found on your website. He had the smoothie four or five days a week and loved it. I use frozen unsweetened strawberries, orange juice, ground flaxseed, and oat bran.

In June I asked the pediatrician if he would order blood work so I could see if the smoothie had helped raise my son's HDL and improved his other blood work. It did. Now his HDL is 34, and his total cholesterol is 138, with LDL of 92. That makes the ratio of LDL to HDL cholesterol much better—under three. It would be great to get his HDL even higher. Do you have any ideas?

A A surprising range of dietary choices can help improve cholesterol profiles, from beets to cinnamon to fish oil to a low-carb diet. The original smoothie recipe you found on our website is: Put orange juice (8 ounces) and a diced banana or peach into a 16-ounce jar, cap the jar tightly, and shake. Add one-third cup raw rolled oats and one tablespoon ground flaxseed meal. Cap once again, shake, and let sit for 15 or 20 minutes. The smoothie can be frozen and will stay cool for hours after coming out of the freezer.

VINEGAR

Q Your column often covers high-cholesterol issues. Why don't you mention the value of adding a daily dose of organic apple cider vinegar as a great way of reducing cholesterol? I add one to two teaspoons to my morning cranberry and orange juice, and my cholesterol has come down from 184 to 132. It's tasty and a whole lot cheaper and safer than the medicines the pharmaceutical industry pushes on us.

Apple cider vinegar is a traditional remedy often suggested for lowering cholesterol. A study in Japan has shown that acetic acid (vinegar) added to the diet can lower cholesterol and triglycerides in rats.[2] We have not seen clinical trials on the effects of vinegar in humans, however.

WALNUTS

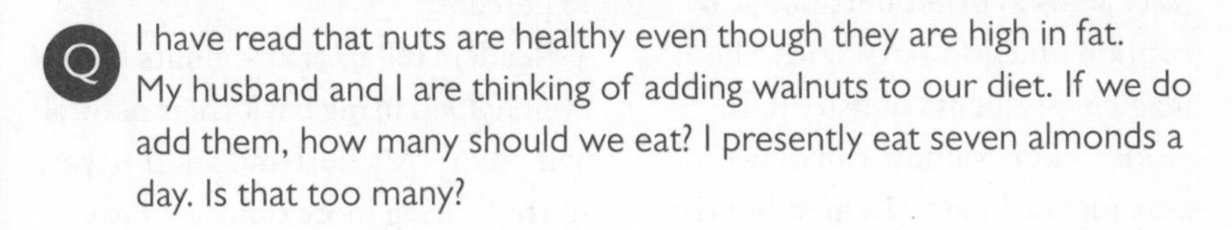

I have read that nuts are healthy even though they are high in fat. My husband and I are thinking of adding walnuts to our diet. If we do add them, how many should we eat? I presently eat seven almonds a day. Is that too many?

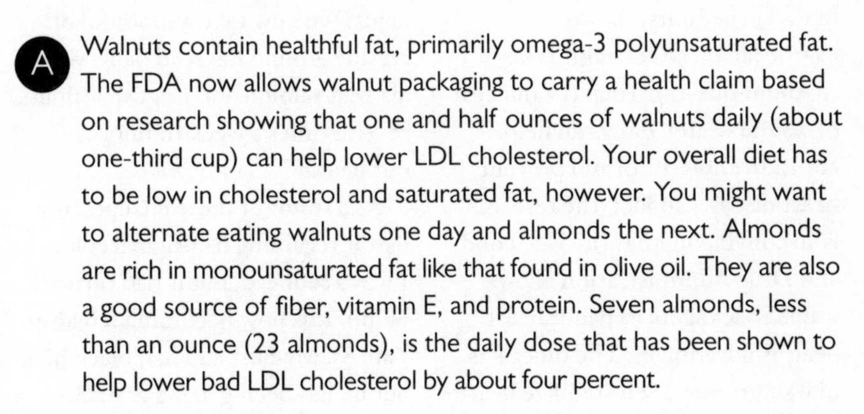

Walnuts contain healthful fat, primarily omega-3 polyunsaturated fat. The FDA now allows walnut packaging to carry a health claim based on research showing that one and half ounces of walnuts daily (about one-third cup) can help lower LDL cholesterol. Your overall diet has to be low in cholesterol and saturated fat, however. You might want to alternate eating walnuts one day and almonds the next. Almonds are rich in monounsaturated fat like that found in olive oil. They are also a good source of fiber, vitamin E, and protein. Seven almonds, less than an ounce (23 almonds), is the daily dose that has been shown to help lower bad LDL cholesterol by about four percent.

Hot Flashes

Hot flashes may be one of the most uncomfortable aspects of menopause. It is embarrassing to suddenly be drenched in sweat during the day. Waking up in the middle of the night frequently interferes with the ability to get a good night's sleep. The standard medical prescriptions for this problem

THE PEOPLE'S PHARMACY

Favorite Food #19: Walnuts

We are nuts about nuts. And while nearly all nuts offer great health benefits because of their monounsaturated fatty acids, walnuts are king.

Walnuts not only have the same fatty acids as other nuts but also contain omega-3 fatty acids, the magic ingredients in fish oil. In moderation, walnuts can do wonders for the heart. (Remember that like all other nuts, they're calorie laden, so restraint is key.)

About one-third cup (1.5 ounces) of walnuts eaten daily can help cut LDL cholesterol and prevent heart disease. In fact, the research is so convincing that the U.S. Food and Drug Administration allows walnut packagers to flaunt heart-healthy advantages. The omega-3s in walnuts may even stabilize heart rhythms. The Physicians Health Study, which followed 22,000 male doctors for nearly 20 years, showed that subjects who ate nuts were considerably less likely to drop dead of a sudden heart attack.[1]

In addition to benefiting the heart, walnuts also appear to cut the risk of developing type 2 diabetes. In a major study published by nutritional expert Walter Willett and his colleagues, women who ate about five ounces of walnuts each week reduced their chances of getting type 2 diabetes by an impressive 27 percent.[2]

Readers tell us that walnuts have even helped bring back their natural hair color: "My husband and I have started eating more omega-3 fatty acids. We now take a spoonful of freshly ground flaxseed daily. We also eat salmon and English walnuts several times a week to improve our health.

As a result of these changes, my hair is regaining its original color. I was a redhead, but it had turned blond. It is now becoming red again. I am 85. My husband had black hair, but he has been getting gray. His hair is now showing more black. He is 86."

We cannot promise that walnuts will restore hair color or act as a fountain of youth. But adding an ounce or so of walnuts to your daily diet will certainly be good for you and may act as a preventive for heart disease and type 2 diabetes.

are estrogen (Estrace, Premarin, Prempro) and antidepressants like fluoxetine (Prozac) or venlafaxine (Effexor). Although such drugs ease hot flashes while you take them, the hot flashes may recur when you discontinue the medications. These drugs all have side effects that may be troubling. For example, estrogen and progestin have been linked to breast cancer. That's why many women are interested in home remedies.

BLACK COHOSH AND ST. JOHN'S WORT

Q I know the experts say black cohosh is ineffective for hot flashes, but personally I wouldn't be without it! I am not one to take supplements without good reason, so periodically I stop taking them to see if they really make a difference. Within 24 hours of ceasing black cohosh, sizzling hot flashes are back with a vengeance! As soon as I restart the capsules, the problem ceases. I wonder if any of the experts have experienced hot flashes themselves. Maybe this would affect their outlook.

A One study showed that a standardized black cohosh extract offered no benefit over a placebo for symptoms of menopause.[1] Other randomized studies support your observation that black cohosh is helpful to relieve hot flashes.

Q I have been taking St. John's wort and black cohosh to relieve symptoms of hot flashes and night sweats associated with menopause. This combination works really well. Are there any negative effects that should concern me?

A The combination of St. John's wort and black cohosh is often used in Europe to treat menopausal symptoms. There are, however, a few potential pitfalls. Black cohosh has been linked to rare cases of elevated liver enzymes. You should ask your doctor to monitor yours. St. John's wort can interact with many medications. Do not take it with any drugs unless you verify with your pharmacist that there is no interaction. More disconcerting is the possibility of eye damage. St. John's wort contains hypericin. When this compound is exposed to visible light it becomes activated and creates dangerous chemicals called free radicals. Researchers at Fordham University have found that hypericin can harm lens tissue and might also damage the retina.[2] Anyone taking St. John's wort for depression or menopausal symptoms should avoid sunlight and even bright indoor light. Sunglasses cannot protect the eye adequately against this possible side effect.

BORAGE SEED OIL

Q You recently had a question from a woman suffering from hot flashes due to menopause. I am a cancer patient, and I avoid soy because it acts like estrogen. I have found that borage seed oil nearly eliminates my hot flashes. The cancer center checked on it before I began and said it's okay.

A You were smart to have the cancer center check your supplement first. Borage seed oil is rich in gamma-linolenic acid, similar to the fat found in evening primrose oil or flaxseed oil. Although the Web has many sites that recommend borage seed oil for hot flashes, we were unable to find a definitive study showing that this dietary supplement can reduce them. Make sure that your supplement has had the pyrrolizidine alkaloids removed, since these can be toxic to the liver.

PYCNOGENOL

Q Help! My menopausal hot flashes are becoming unbearable and debilitating. I have tried many remedies. Some (like cutting down on caffeine) helped a little, but others (like soy) did nothing. I work with liver transplant patients, and specialists say that the herb black cohosh can damage the liver. So I'm afraid to try it. I finally broke down and tried an estrogen patch my doctor prescribed. I had an adverse reaction to it in less than a week. What can you recommend?

A Although there are reports of liver problems associated with black cohosh, this appears to be an uncommon complication. One study suggests that a patented pine bark extract can help ease hot flashes and other symptoms of menopause. The compound is Pycnogenol, derived from the French maritime pine. The study included 155 women aged 45 to 55. After six months of treatment with Pycnogenol or a placebo, women treated with the pine bark extract had significantly fewer symptoms and lower cholesterol levels than those taking the placebo.[3] It may be worth a try.

Q I was skeptical about Pycnogenol, but after four days of use, my severe hot flashes have completely disappeared. I take 50 milligrams in the morning and 50 milligrams again in the evening. I used to have a couple of hot flashes every hour around the clock. Since I have started taking Pycnogenol, I haven't had one. It feels like a miracle!

A We have seen just one controlled study of the effects of this French pine bark extract on hot flashes.[4] It looks promising. We are glad it helped you. It can be found on the Web or in health food stores.

Q I have a condition (Wegener's granulomatosis) that made me miserable for years and stumped several doctors. The symptom that troubled me was like hot flashes, but instead of flashing they were almost constant. I finally searched "hot flashes" on *www.peoplespharmacy.com*. Now they are gone, and I'm telling everyone! Taking 50 milligrams of Pycnogenol twice a day has done the trick. Thank you so much. I haven't had a flash since the day I started it. My disease is not curable, but the flashes bothered me more, and they are gone.

A We are pleased that Pycnogenol worked for you. The study we cited showed it helps reduce hot flashes for menopausal women.[5] We did not know that it might work for hot flashes from other causes.

YAMS

Q I am 52. At age 49 I began to have menopausal symptoms—irregular periods, hot flashes, night sweats, depression (not wanting to get out of bed in the morning), fuzzy thinking, and vaginal dryness that made intercourse painful. I have a very healthy lifestyle including a vegan diet, daily exercise, no smoking, and almost no alcohol. I've always believed that the right nutrition allows the body to handle anything. But menopause really threw me for a loop. Somewhere I'd heard that yams could help support hormones. So I began baking yams and eating some every day. I was certainly a skeptic. However, after only five days of eating yams (one-half per day, depending on size), I stopped having hot flashes and night sweats altogether! Within days I realized that the vaginal dryness problem was gone. My thinking had cleared, and my depression began to lift. In addition, my breasts have increased in size and feel full instead of saggy and droopy. My normal menstrual periods have returned. What surprising results I've gotten from a simple (and delicious) food. The key is to eat them daily.

Research suggests there does appear to be an estrogenic effect from regular yam consumption.[6] Taiwanese investigators fed post-menopausal women yams (*Dioscorea alata*) for 30 days and noted improvements in hormone and cholesterol levels. Control subjects were fed sweet potatoes and did not get similar benefits. Investigators concluded that changes brought about by consumption of yams "might reduce the risk of breast cancer and cardiovascular disease in postmenopausal women." In the United States, sweet potatoes and yams are often confused, but they are completely different plants.

Incontinence

One of the most distressing conditions at any age is incontinence. Involuntary leakage of urine is embarrassing and directly affects quality of life. A urologist should always investigate this problem to uncover any underlying medical conditions that might be treatable. Beware of some prescription drugs that are promoted for this problem or for "over active bladder," as they may cause disorientation, drowsiness, or even hallucinations.

DIETARY CHANGES

Q I was diagnosed with an overactive bladder six months ago. The urologist prescribed VESIcare and handed me a pamphlet on the main offenders that cause the symptoms. After reading the pamphlet, I eliminated caffeine, chocolate, and alcohol from my diet. Voila, I make no more nightly visits to the bathroom. As an added benefit, I no longer need any medication for reflux. Sometimes giving up something that is harmful to health is better and less expensive than taking medicine.

A Not everyone will get as much benefit from the kinds of changes you made, but this is a low-risk approach. Chocolate, caffeine, and alcohol all have been implicated as possible culprits in triggering reflux as well.

KEGEL EXERCISES

Q I am 29 years old and engaged to be married this summer. I am very excited, but as we make plans for our honeymoon, I am starting to get worried as well. And I'm too embarrassed about this problem to know where to turn for help. I am in good health, but sometimes when I cough or sneeze, I can't hold my urine in. This is bad enough, but every so often at night I dream that I am looking for a restroom, and when I wake up the bed is wet. I would be mortified if this happened on my honeymoon. Isn't there some medication children take to keep them from wetting the bed? Would it work for me too?

A Make an appointment with a physician for proper diagnosis. Bladder training and special exercises to strengthen the pelvic muscles (Kegels) can be very helpful for your case of stress incontinence. Some doctors may also prescribe estrogen or oral decongestants to tighten the sphincter. To keep from wetting the bed at night, avoid caffeine during the day and reduce fluid intake a few hours before bedtime. If this doesn't help, the urologist may consider prescribing desmopressin (DDAVP), a nasal spray more often used for children. This medication dramatically reduces urine formation and the chance of an accident.

STINGING NETTLE

Q I have a urinary drip that used to require wearing heavy pads. My doctor prescribed Oxytrol and then Detrol. Both medicines made my eyes, mouth, and throat unbearably dry. I started taking stinging nettle for allergies and postnasal drip. I found it very helpful. In addition, I no longer need to wear pads, only panty liners, because it helped my urinary problem too.

A The herb stinging nettle (*Urtica dioica*) has been used in Europe to relieve allergy symptoms and improve urinary flow in cases of benign prostate enlargement. We are pleased it helped you. Some people may be allergic to nettles. If a rash develops, you should discontinue the herb. Stinging nettle should not be used during pregnancy. New research suggests that drugs like Detrol, Ditropan, and Oxytrol, which dry out mucous membranes, may also impair mental function.

Insomnia

Grandma had it right. We all need a good night's sleep. Sleep is important for immune function, memory, and general well-being. But one American out of five suffers from insomnia. Chronic insomnia is linked to weight gain, high blood pressure, diabetes, poor cognitive function, reduced daytime performance, and falls. Ruling out contributing factors, such as sleep apnea or another medical condition, requires consultation with a sleep specialist. Sleeping pills may help, but they can be habit forming and may have dangerous side effects. Some readers report instances of sleep driving that are associated with the prescription sleep aid Ambien. If a hot bath to reset your body temperature or a small high-carb snack before bedtime doesn't help, there are some other home remedies that might make a difference for you.

ACUPRESSURE

Q You wrote about a person who had taped a bean to the inside of his wrist at bedtime as an aid against insomnia. Pushing on an acupressure point gave him a decent night's sleep. Would you use the same bean over and over or a new bean every night?

A That reader told us that he taped a dried kidney bean between the two tendons on the inside of his right wrist. He located a spot three finger widths from the crease of the wrist between the two tendons. According to our research, this is an acupressure point called the inner gate. Pushing on it is supposed to relieve anxiety and promote sleep. You should be able to use a dried kidney bean many times without replacing it. You may also want to look for acupressure wristbands, with a plastic button embedded in an elastic or Velcro strap. The wristbands are available online or in drugstores, at a wide range of prices.

Q Thanks for your tip on the acupressure sleep aid. I use a Sea-Band wrist strap that improves my sleep and also helps me fall back to sleep when I wake up. But the best thing is that it reduces my snoring. It's so effective in reducing snoring that when I forget to wear it, my wife wakes me up so I can put it on. I even use it for a nap, so my snoring won't wake me up. I think the reduced snoring effect is a sleep aid in itself.

A Sea-Bands are intended to prevent motion sickness. They press on an acupressure point on the inner wrist. The few studies we found on sleep and acupressure were conducted in Korea and Taiwan. We're glad to hear Sea-Bands helped your snoring as well as your insomnia.

MAGNESIUM

Q I often read about people having trouble getting a good night's sleep. Perhaps my story will help someone. My pharmacist recently told me to take my magnesium tablets at bedtime instead of with breakfast. What a difference in falling asleep!

A Another reader shares your enthusiasm for this mineral: "Magnesium is like a tranquilizer for me. I enjoy deep sleep like when I was a teenager now that I take a hot magnesium citrate tea called Natural Calm before bed." Although other readers have also reported benefits, we could find no scientific support for the claim that magnesium eases insomnia. Too much can cause diarrhea. People with kidney problems should avoid magnesium.

TEA

Q Your readers often ask about help with falling asleep easily. It's a good thing I was safely seated on my own sofa when I first sipped a cup of Tazo Calm herbal infusion tea. I was soon napping comfortably. I now use this blend when I want to relax. It is difficult not to doze off. I'm not pushing the Tazo brand. The blend contains chamomile, hibiscus, spearmint, lemongrass, rose petals, blackberry, sarsaparilla, lemon balm, licorice, and natural flavors. Chamomile tea doesn't put me to sleep. I leave it to you to figure out the magic.

A Thanks for the tip. Other readers will surely want to try this tasty blend. Chamomile is traditionally considered a mild sleep aid. Perhaps the secret of this tea is in the combination of other mildly relaxing herbs, such as lemon balm. Hops, valerian, and passionflower are other herbs used for relaxation and sleep.

Irritable Bowel Syndrome

Irritable bowel syndrome (IBS) is a mystery. Cause and cure remain elusive. Some people complain of diarrhea, while others suffer from constipation or an alternating cycle of both. Other symptoms can include bloating and abdominal pain. Your doctor should rule out celiac disease, an autoimmune disorder brought on by exposure to wheat, barley, and rye, before making a diagnosis of IBS.

COCONUT

Q Thank you so much for writing about IBS and coconut macaroon cookies. They work! I suffered from chronic diarrhea for years and have been healed for the last two years.

A We're always pleased to hear about success with home remedies. Donald Agar wrote us nearly ten years ago to report that two Archway coconut macaroon cookies a day banished the chronic diarrhea he suffered as a consequence of Crohn's disease. We have heard from other readers that coconut helps combat diarrhea. You can read stories from people who have tried it at *www.peoplespharmacy.com*. Not everyone who suffers from IBS benefits from coconut macaroons, however.

Q I couldn't find Archway coconut macaroon cookies at the supermarket, so I made my own. Presto! I am normal. It is still a miracle to me. I wonder if it isn't something in the coconut that is the key. After all, cookie ingredients are about the same, commercial or homemade. I ate two a day for a week. Just a couple a week is not enough for me. It is hard to believe there is some relief for this awful condition.

A We think there is something about coconut that may provide relief for some people. One reader also had trouble locating Archway brand cookies and reported, "I just started eating some shredded coconut. The symptoms went away after the first day."

PEPPERMINT

Q I've heard that taking enteric-coated peppermint oil can help IBS. My worst symptoms are stomach cramps and flatulence, which make me reluctant to eat out or travel. Some of my friendships have suffered. What does enteric mean, and why would peppermint oil help?

A Irritable bowel syndrome can produce symptoms such as abdominal pain, gas, cycles of constipation and diarrhea, urgency of stools, and bloating. Enteric-coated peppermint oil has been shown to relieve symptoms of IBS. One study showed that this preparation reduced bloating, abdominal pain, flatulence, frequency of bowel movements, and stomach noises.[1] A German investigator reviewed 16 clinical trials in which enteric-coated peppermint oil was used to treat symptoms of IBS. Two-thirds of the placebo-controlled trials indicated that peppermint oil eased symptoms twice as well as a placebo.[2] Peppermint in the stomach could make heartburn worse, which is why enteric-coated pills are essential. This special coating, by definition, doesn't dissolve until it gets to the small intestine. There, peppermint oil eases spasms and relieves symptoms with relatively few side effects. Some people have reported heartburn and rectal burning, however. One brand-name product, Pepogest, can be bought online.

Q Is there a home remedy to ease the discomfort of IBS? I currently use Nexium daily and drink a lot of water, but I am always uncomfortable. I'm at my wit's end. HELP!

A Research from Germany suggests you may want to try enteric-coated peppermint oil. There were some side effects: Peppermint oil can cause heartburn and rectal burning. But overall it seemed the enteric-coated capsules improved quality of life for people with IBS.

PROBIOTICS

Q My teenage daughter has been taking antibiotics to treat her acne for years, but she's also had terrible gastrointestinal problems (stomachaches and diarrhea) for much of that time. I didn't think of a connection until recently, but now I wonder if the antibiotics might be responsible. She has taken Prilosec per her doctor's recommendation, but it really hasn't helped. Is there anything else that might help her overcome these symptoms? Her dermatologist says if she stops the minocycline she is taking her acne will come back badly, and I hate for her to have to deal with that at the start of the school year.

A It is possible that years of antibiotic treatment have altered the ecology of your daughter's digestive tract and contributed to her pain and diarrhea. Antibiotics kill good bacteria as well as bad ones. Repopulating the digestive tract with good bacteria can sometimes help reverse that problem. Such probiotic bacteria may be found in yogurt with active live cultures or in capsules such as Culturelle, Enzymatic Therapy, or Florastor.

Q Are there any natural treatments for IBS? I think they should call it cranky colon or irritable intestine or something alliterative, but I've had it for several years and it seems that there is little to be done for it. I have occasionally had rectal spasms so intense I pass out. Doctors don't have anything to offer. Do you know of anything I could try?

A One study showed that a probiotic product containing *Bifidobacterium infantis* was significantly better than a placebo for constipation and diarrhea associated with IBS. A Procter & Gamble product called Bifantis was used in the research.

Q My husband and I travel by car each year to Arizona for the winter and return home in spring. The trip is difficult because I have IBS and often hold up our fellow travelers so I can get to a bathroom. At present I take Imodium, but that only works after a bout of diarrhea. Is there anything I could take to make our trip less stressful? It does not feel good to drive through the mountains with no rest stations on the horizon when my stomach is "erupting."

A We don't have any magic bullets to offer you, but we have heard from several readers that a probiotic product called Digestive Advantage Irritable Bowel Syndrome was helpful. It contains live strains of *Lactobacillus* bacteria. Here is what one reader had to say: "For years I have suffered with nausea and gastritis. Antacids and other stomach medicines did not help. Then my doctor told me his daughter has the same problem. A product called Digestive Advantage Irritable Bowel Syndrome helps her. It costs under ten dollars at most drugstores. This product works fast, and I have been totally well for months."

TURMERIC

Q Having read a few of your articles on the benefits of curry powder, I would like to add that it helps irritable bowel syndrome. After having suffered with IBS pain for several years, I looked into possible natural remedies and found that naturopaths recommend curry for certain intestinal problems. Since using curry in my food once a day, I have been pain free. The bloating is still there, but it doesn't hurt. If I don't sprinkle curry powder on my food for a few days, the pain returns.

A Curry powder contains turmeric, a yellow spice that has anti-inflammatory properties. Others have told us that it eases their arthritis pain and psoriasis symptoms as well. We hadn't heard that it might relieve symptoms of IBS.

Joint Pain and Arthritis

As many as 70 million Americans suffer from the chronic joint pain that we call arthritis. Of the many varieties, the most common is osteoarthritis, which tends to develop with age. Why some people suffer from stiff, sore, aching joints and others are seemingly immune from arthritis remains a mystery. The drugs that are most often used against arthritis pain—the nonsteroidal anti-inflammatory drugs (NSAIDs) like ibuprofen or diclofenac—may often ease pain in the short term. However, side effects, such as digestive tract irritation, including bleeding ulcers, as well as possible increased risk of hypertension, heart attacks, stroke, or congestive heart failure, certainly limit the appeal of these drugs. That's why inexpensive home remedies for joint pain and arthritis are so popular.

BEE STINGS

Q I'm a nurse in a rural hospital. Some of the mountain folk I care for tell me that a bee sting every two years or so will significantly decrease arthritis inflammation and pain. They attribute awareness of this remedy to the Chinese who came to this area a hundred years ago to work on the railroads and in the logging industry.

A Apitherapy, or bee venom therapy, for arthritis dates to ancient Egypt and China. Hippocrates (460–377 B.C.) is said to have written about bee stings for treating painful joints. Doctors in this country used apitherapy to treat arthritis during the first part of the 20th century. Hospital pharmacies even stocked venom for injections. After World War II, this approach was considered unscientific and fell out of favor. Proponents claim that honeybee stings can alleviate the pain of tendinitis, arthritis, multiple sclerosis, and postherpetic neuralgia (nerve pain that lingers after a shingles attack). The American Apitherapy Society can provide more information (*www.apitherapy.org*).

Q I was stung on my left leg five times by yellow jackets. I have osteoarthritis in my left knee, and the pain has been gone since I was stung. I'm hoping that it will last!

A You're not the first person to share such a story with us. Years ago a reader wrote, "While snoozing on the porch I was stung on the finger by a tiny bee. The result: intense pain and, after that, a great reduction of arthritis in my arm." Today apitherapy is undergoing a resurgence. Proponents claim that honeybee stings can alleviate the pain of arthritis. Yellow jackets can be dangerous, however, and should not be used in apitherapy. People who are allergic to bee stings should never try this therapy, which is best applied by a trained apitherapist.

BOSWELLIA

Q I have arthritis in my fingers. Using the computer compounds the pain. I can't take anti-inflammatory medication because I have an ulcer. Could you suggest other supplements that might help?

Many herbs and dietary supplements can ease inflammation. One person offered the following: "The combination of boswellia and glucosamine-MSM replaces nonsteroidal pain relievers and works well for me. Nine years ago I was literally falling down because of the pain in my spine. I heard someone say that the herb boswellia had saved her life because of back pain. That Saturday I started boswellia. In two weeks the pain decreased, and after a month there was an enormous difference. At last I could sleep and I could walk. Several years later the arthritis increased and I added MSM [methylsulfonylmethane supplement], and glucosamine and chondroitin. I take them with boswellia and get good relief."

CAYENNE

Q You recently had a question from a person with arthritic fingers who can't use anti-inflammatory drugs because of an ulcer. Many years ago, an old man panning gold in icy water told me of an arthritis cure. He'd had arthritis in his fingers so bad he could hardly move them. He started taking a teaspoon of cayenne a day in a small glass of tomato juice. He said this remedy took a month to take effect and a month to wear off. When I got arthritis in my hip, I started taking cayenne. I found that one-quarter teaspoon a day in a tall glass of orange juice works for me.

A We've heard lots of arthritis remedies over the years but never one recommending cayenne. It is, however, a time-honored ingredient in arthritis rubs.

CERTO AND JUICE

Q I read in your column that someone used Certo in juice to relieve painful arthritis in the hands. Is this the pectin with which one makes jelly? I did buy some and started putting a tablespoon in eight ounces of pomegranate juice, as I have horrendous arthritis in my hands.

A Certo is a liquid pectin product used by home canners to make jams and jellies thicker. For ten years readers have told us that Certo mixed with grape juice can help ease arthritis pain. One formula calls for a tablespoon of Certo in eight ounces of juice daily.

Q I read about grape juice and Certo last fall while I was suffering a painful medial column collapse of the right foot that the doctor attributed to psoriatic arthritis. I was going to have surgery in late November, but I tried the juice with Certo in October. I didn't tell the doctor, as I didn't want to be admonished. For some mysterious reason, the pain subsided. The doctor was amazed but said surgery should be done only as a last resort. I see him next month for a checkup and will mention the juice with Certo. I'm not saying that's what put the arthritis in remission, but something happened, as I could hardly walk at the time. So I'll continue taking it.

Q I've read your columns about grape juice and Certo for easing joint pain, but I don't like grape juice. I tried pomegranate juice with Certo instead. It's much lower in calories, and it tastes really good. After just a couple of days, the results are amazing! I can comfortably make a tight fist now, which means I can look forward to throwing punches in aikido classes again without jamming an arthritic knuckle. As an emergency medical technician (EMT), I'm in tune with conventional modern U.S. medicine. From a medical standpoint, this remedy has me stumped, but there is no doubt that it worked quickly and effectively for me. I don't know if this mixture will affect other drugs or conditions, so others should consult their doctors before trying it.

A Thanks for sharing your experiments. We have heard from hundreds of readers that Certo and grape juice can ease joint pain. Research suggests that pomegranate juice also can ease inflammation and slow cartilage destruction.[1]

CHERRY JUICE

Q My husband and I take black cherry juice concentrate for arthritis aches and pains. I buy it at the local health food store. We take a teaspoon a day, like cough syrup. My finger joints are no longer swollen and painful. On rare days when I still have discomfort, I just take another dose.

A Tart cherries, sour cherries, and black cherries have all been used to combat inflammation associated with arthritis or gout. Cherry juice concentrate is usually more affordable than fresh cherries or juice. It can be added to seltzer water or made into a tea. Concentrated cherry capsules or cherry supplement bars are also available. Studies on animals have shown that the red anthocyanin compounds in cherries act as anti-inflammatories.[2] Research from the 1950s also suggests that cherries may help both gout and arthritis.

Q I would like some info about cherries for a friend who suffers from arthritis and gout.

A One study suggested that sweet cherries could lower uric acid levels that cause gout. The researchers also observed that "cherries may inhibit inflammatory pathways."[3] Readers have been telling us that sour cherries (fresh, dried, frozen, juice, or CherryFlex pills) can ease both gout attacks and arthritis pain. Here is one recent account: "I tried the cherry juice after reading about it in your column. I now get out of bed without back pain (after less than a month). I have sciatica, along with other back issues, and arthritis in both knees. I can do stairs again. On top of that, I asked our vet if I could give it to our dog. He said yes, but to give her half a dose. I give her two capsules a day, and she is now running again after only two weeks."

CURCUMIN/TURMERIC

Q All my life my knees have ached at night. I used Aleve, arthritis-strength aspirin, or Tylenol and usually woke up about 3 a.m. and took more. I read in your column about using turmeric for arthritis pain and bought some capsules. I took one with milk and a cookie at bedtime and slept pain free all night—and every night since then. It is almost miraculous.

THE PEOPLE'S PHARMACY

Favorite Food #20: Cherries

There's a reason for the adage "Life is just a bowl of cherries." But probably not even Ethel Merman, when she sang the famous song with this title back in 1931, could have guessed at all of the remarkable and life-enhancing properties of this delectable fruit.

Cherries can literally put the spring back in your step. Research indicates that people who suffer from arthritis or plantar fasciitis may benefit considerably from the proven anti-inflammatory properties of cherries.[1] A study from several years ago found that as little as ten ounces of cherries a day decreased inflammatory markers in the blood.[2]

Tart cherries contain anthocyanin compounds that inhibit enzymes called COX-1 and COX-2. These are the enzymes targeted by anti-inflammatory drugs like Celebrex, diclofenac, and ibuprofen, so it is not surprising that cherry juice appears to alleviate pain.[3] One reader who suffered from arthritis pain in the hip and knees reported, "I took tart cherry juice every morning and saw results after the third week. I have used tart cherry juice for over one year now, and I am pain free!"

Sweet cherries also appear to reduce the high uric acid levels that cause gout,[4] and they have been shown to be quite effective at relieving pain caused by this often agonizing condition. We've also heard some anecdotal evidence that cherries may help fibromyalgia sufferers soothe discomfort and pain.

For people who don't love the taste of cherries or for those who are concerned about the high sugar and calorie content of juice, it is also possible to buy cherries in the form of supplements. Two companies that sell capsules are Brownwood Acres, which sells CherryFlex, available at *www.cherryflex.com,* and Fruit Advantage, available at *www.fruitadvantage.com* or at 877-746-7477.

The next time you're thinking about popping a pain pill for sore joints or for aching feet, consider trying cherry capsules instead. You may find some sweet relief.

We are intrigued by your story. Studies show that turmeric has anti-inflammatory properties. We have heard from readers that it may increase the risk of bleeding in combination with Coumadin, an anticoagulant used as a blood thinner.

EQUAL SWEETENER

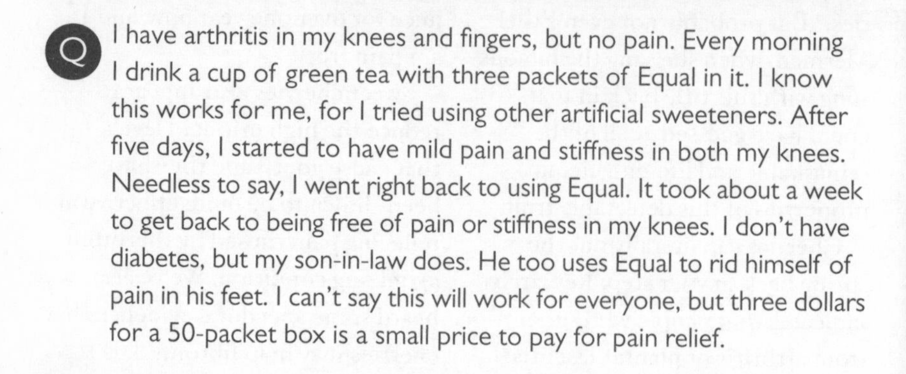

Q I have arthritis in my knees and fingers, but no pain. Every morning I drink a cup of green tea with three packets of Equal in it. I know this works for me, for I tried using other artificial sweeteners. After five days, I started to have mild pain and stiffness in both my knees. Needless to say, I went right back to using Equal. It took about a week to get back to being free of pain or stiffness in my knees. I don't have diabetes, but my son-in-law does. He too uses Equal to rid himself of pain in his feet. I can't say this will work for everyone, but three dollars for a 50-packet box is a small price to pay for pain relief.

A Aspartame (Equal) has been tested for arthritis pain in a small, placebo-controlled trial. It is nearly as effective as aspirin but does not irritate the stomach. People who cannot tolerate aspartame or prefer to avoid it may want to try natural anti-inflammatory agents such as ginger, curcumin, boswellia, or bromelain.

FISH OIL

Q I have heard that the omega-3 fatty acids in fish oil may be beneficial in helping the body repair cartilage and improving arthritic joints. Is this an old wives' tale, or does it hold up under scrutiny?

Q Fish oil is amazing. My husband is under a doctor's care for arthritis. Within one day of starting fish oil, the swelling began to go down in his fingers. He still takes his prescribed medication, but in lower doses. This relief is so much better.

A There have been a number of studies suggesting that fish oil is helpful for joints affected by rheumatoid arthritis. Even when doctors do not detect an objective difference, patients taking fish oil report less pain and morning stiffness and take fewer pain relievers than patients who refrain.[4]

GIN AND RAISINS

Q A neighbor of mine was so crippled with arthritis that she had to use a walker. She and her husband had over 100 beautiful azaleas that she could no longer care for. Then I walked by her house and thought I was seeing things. There she was, down on her knees, working in her flower beds. I said, "Nancy, have you experienced a miracle?" "No," she said, "just gin and raisins!" I began taking the recipe and was able to stop going to the arthritis clinic, which wasn't helping anyway. My osteoarthritis subsided, and I stopped the remedy. Now, ten years later, I'm 67 and the arthritis is back in my little fingers, with redness, pain, and swelling. I remembered the recipe but forgot to let the gin evaporate. It turned into the most delicious raisiny brandy and helped my joints as well. In the next batch I used black raisins. For some reason, it isn't helping as much.

A We cannot begin to explain why some folks benefit so dramatically from the gin and raisin remedy while others tell us it is worthless. Nor can we explain why light raisins might work better than dark ones in some cases.

Q I am interested in the formula for the gin and raisin remedy for arthritis. How much of each makes a batch, and how many batches would I take each day?

A This recipe calls for golden raisins. Empty a box of raisins into a shallow bowl, and then add enough gin to cover them. Allow the gin to evaporate. (This may take up to a week.) Store the raisins in a covered container, and eat nine golden raisins daily. Although we don't know why, this remedy works for many people. But a word of warning: People who are allergic to sulfites should avoid golden raisins, which have sulfites added to preserve their color. One person experienced a swollen tongue due to this allergy.

HONEY AND VINEGAR

Q About a year ago you mentioned a recipe for arthritis from a football player. He mixed one-half teaspoon apple cider vinegar and one-half teaspoon honey in a glass of water with a teaspoon of orange powder. I'd really like to try this remedy, but I cannot find orange powder. Did he mean Tang?

A The reader who gave us this recipe got it from a former owner of the Dallas Cowboys. The mysterious orange powder is nothing more than orange-flavored Knox gelatin. At age 81, he says this formula still keeps his knuckles flexible.

MAGNETS

Q My husband says that his magnetic bracelet helps his arthritis pain. What do you have to say about that? Would it also help my fibromyalgia? Are there side effects?

A A few randomized studies suggest that magnets may help arthritis pain. Many scientists are skeptical, and after one small, placebo-controlled trial, researchers concluded that magnetic and copper bracelets are useless for treating arthritis pain.[5] Some people who have tried magnets agree with your husband, however. Magnets are not appropriate for pregnant women, people wearing pacemakers, or people using electromagnetic equipment such as insulin pumps or sleep apnea machines. There are no studies to show whether a magnetic bracelet would help people with fibromyalgia, although one reader reported that a magnetic bracelet made her fibromyalgia pain worse.

PINEAPPLE JUICE

Q Years ago, you wrote about an enzyme in pineapple juice that helps with arthritis pain. At that time I was in my early 40s and was already having pain in my hands and feet from arthritis. I started drinking one glass of pineapple juice a day, and my symptoms cleared up. I may eventually develop arthritis, but hopefully it will not be as severe as it would have been.

Pineapple juice contains bromelain, which appears to have anti-inflammatory properties. In one study, a product containing bromelain (Phlogenzym) was effective in easing discomfort from hip arthritis.[6]

PLANT PECTIN AND CERTO

Q I woke up this morning to find I was another year older, but thanks to you, I am active again for the first time in years. I combined several suggestions I found in your column. I am taking glucosamine and chondroitin along with grape juice and Certo (plant pectin) and turmeric. These remedies made me feel so good I forgot to take them for a few days, and the pain returned. I won't make that mistake again.

A **The experiment you are conducting is unusual. Combining several natural remedies for arthritis hasn't been tested but may offer some advantages.**

SLOE GIN

Q Golden raisins soaked in gin were ineffective against my arthritis pain. But then I tried raisins in sloe gin, and they were immediately and totally effective.

A **Regular gin is flavored with juniper berries, while sloe gin is flavored with sloe berries from the blackthorn bush, traditionally used for digestive disorders. This isn't the first time we have heard that sloe gin with raisins may fight arthritis pain.**

SOAP

Q My daughter heard that a bar of soap at the foot of the bed between the sheets would ease arthritis pain. I tried it. After about four weeks my arthritis seems to be much better. Is there something in the soap that helps, or is it my imagination?

We have reported on a home remedy for leg cramps that calls for a bar of soap to be placed beneath the bottom sheet, near the legs. We didn't think about this working for arthritis until we received this note from a reader:

"Since my husband sometimes gets leg cramps, I gave him your article about carrying soap in his pocket. He decided to try it, and for at least four days now he has had no pain from his sciatica. In fact, he has not had to take the pain medication that he usually takes daily. Have you heard of this effect from anyone else? Do you have any conjectures on why a bar of soap works?"

We have no idea why soap might work, but it is not only cheap but also harmless.

VITAMIN D

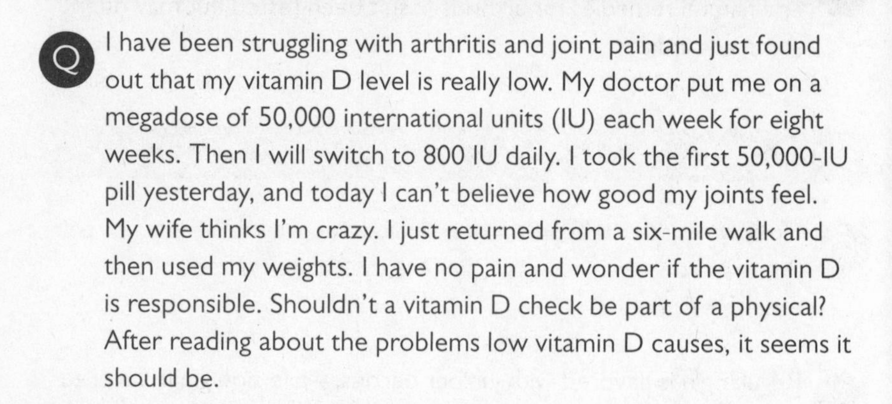

Q I have been struggling with arthritis and joint pain and just found out that my vitamin D level is really low. My doctor put me on a megadose of 50,000 international units (IU) each week for eight weeks. Then I will switch to 800 IU daily. I took the first 50,000-IU pill yesterday, and today I can't believe how good my joints feel. My wife thinks I'm crazy. I just returned from a six-mile walk and then used my weights. I have no pain and wonder if the vitamin D is responsible. Shouldn't a vitamin D check be part of a physical? After reading about the problems low vitamin D causes, it seems it should be.

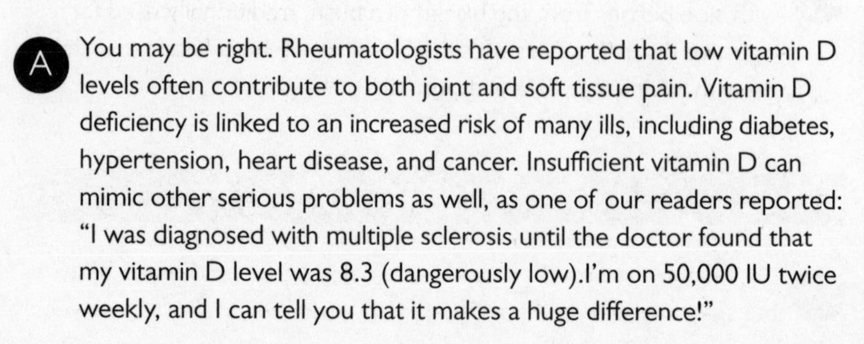

A You may be right. Rheumatologists have reported that low vitamin D levels often contribute to both joint and soft tissue pain. Vitamin D deficiency is linked to an increased risk of many ills, including diabetes, hypertension, heart disease, and cancer. Insufficient vitamin D can mimic other serious problems as well, as one of our readers reported: "I was diagnosed with multiple sclerosis until the doctor found that my vitamin D level was 8.3 (dangerously low).I'm on 50,000 IU twice weekly, and I can tell you that it makes a huge difference!"

Kidney Stones

Words are inadequate to describe the pain of a kidney stone. Women who have gone through childbirth often say that passing a kidney stone is far worse than giving birth. You get the picture. And once you have had your first kidney stone, your chances of getting another one go up significantly. Why some people are so susceptible to kidney stones and others are not is one of the many mysteries to which we have no answer. Here's another puzzle: Why is the southeastern part of the United States the Kidney Stone Belt? The farther north you travel, the less likely you will be to develop a stone. Is it the temperature? (Some folks say global warming will dramatically increase the number of kidney stones in the United States.) Others blame kidney stones on diet—possibly the large amounts of iced tea so popular in the Southeast. Whatever the cause, the goal is to prevent them. Roughly three-fourths of all kidney stones are composed of calcium and oxalate. Stones of calcium phosphate are less common. It may be possible, if you have passed a kidney stone, to discover which type you make. Most experts recommend increasing water intake to keep calcium from concentrating in the urine and forming crystals.

CALCIUM-RICH FOODS

Q Whenever I read about calcium, only women's needs are addressed. Little comment is made about whether men should be concerned about lack of calcium as they age. My husband thinks if he takes calcium supplements, he might get kidney stones. Is this true?

A Although men are less susceptible to osteoporosis than women are, they are not immune. Adequate calcium intake is also important for men. Your husband is correct that calcium pills may increase the risk of kidney stones, but calcium from food seems to protect against this painful condition. If your husband refuses to drink milk, perhaps you could add yogurt, sardines, canned salmon, and other calcium-rich foods to the menu.

LEMONADE AND ORANGE JUICE

Q I have had a couple of kidney stones, and my urologist recommended lemonade to prevent them. I was skeptical until I did some checking on the Web and Medline. I learned that lemon juice is a natural source of citrate. This compound helps prevent the formation of calcium oxalate kidney stones. Some lemonade mixes do not list sodium citrate, but canned concentrate appears to have a lot of real lemon juice, and sodium citrate is listed.

A Doctors prescribe potassium citrate to prevent the recurrence of stones made of calcium oxalate. One study shows that adding lemonade as well seems to provide some added benefit. A trial testing cranberry juice found that it too could help reduce the risk factors[1] for kidney stone formation.[2] This popular beverage reduced oxalate and increased citrate levels in urine.

Q Are any foods or supplements especially bad for people who get kidney stones? I would like to know what to avoid and what would be helpful. I never want to experience the pain of passing a kidney stone again.

A The most important recommendation for avoiding kidney stones is to drink plenty of fluids. But the beverage you choose does make a difference. Data from more than 80,000 women in the Nurses' Health Study indicate that coffee, tea, and even wine are associated with a reduced risk of kidney stones, but grapefruit juice seems to increase the risk.[3] This is a puzzle, because when healthy people in one study drank grapefruit juice, their urine seemed no more susceptible to stones than when they were drinking water.[4] Orange juice and lemonade reduce the chances of developing a stone. They increase citrate in the urine, and that reduces crystallization of calcium oxalate into kidney stones. Urologists who have studied "lemonade therapy" have concluded that it represents a viable alternative for patients who have a hard time tolerating standard treatment.[5] Such an approach calls for one to two quarts of unsweetened or low-sugar lemonade per day.

Lice

Lice do not discriminate on the basis of hygiene, age, or economic status. Children get these scalp parasites because they put their heads close together or share hats and combs. Lice don't jump or fly, but they do crawl and spread easily in close quarters like camp or school. And they have developed resistance to common lice shampoos. That's why many look for home remedies for lice.

CETAPHIL

Q Is there a safe and easy improvement on the method of removing head lice? My wife is a kindergarten teacher, and this annual ritual of treating them is wearing us both out. Her students bring lice from home, and they spread to teachers and other students. Please help!

A There is one novel approach that is both easy and safe. Dampen the hair, coat it with the facial cleanser Cetaphil, and then use a blow-dryer. Cetaphil hardens and forms a barrier that suffocates the lice. Leave the Cetaphil on overnight, and then shampoo it out in the morning.[1]

COCONUT OIL

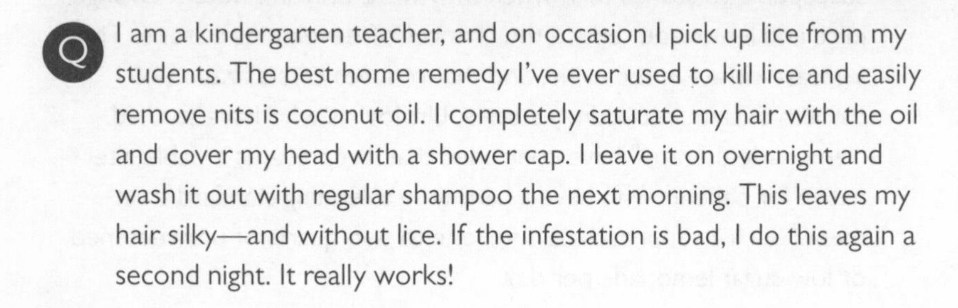

Q I am a kindergarten teacher, and on occasion I pick up lice from my students. The best home remedy I've ever used to kill lice and easily remove nits is coconut oil. I completely saturate my hair with the oil and cover my head with a shower cap. I leave it on overnight and wash it out with regular shampoo the next morning. This leaves my hair silky—and without lice. If the infestation is bad, I do this again a second night. It really works!

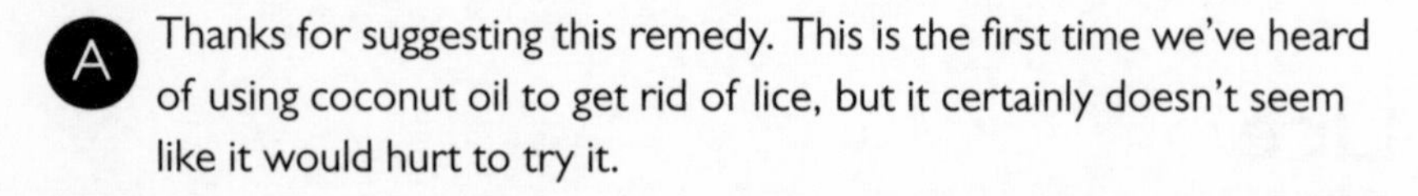

A Thanks for suggesting this remedy. This is the first time we've heard of using coconut oil to get rid of lice, but it certainly doesn't seem like it would hurt to try it.

LISTERINE

Q Both of my boys were sent home from school with head lice. The checklist given to me by the school nurse stated that in order for the boys to return to school, I must treat their scalps with an insecticide. I used a head lice treatment containing permethrin, and it was completely ineffective. Both children were refused readmission to school, and I was instructed to reapply the insecticide that day (despite the package instructions that treatments should be spaced at least seven days apart).

I took them home and washed their hair, towel dried it, saturated it with Listerine, and covered their heads with shower caps. I left the shower caps on for two hours, and then we removed them and I combed their hair with a lice comb. The next day they washed their hair and toweled it dry, and I sprayed their hair with Listerine and combed it again. The lice are gone even though the infestation was severe. Prior to the Listerine treatment I combed hundreds of lice from their hair. Listerine was much more effective than the insecticide.

A There are reports that lice have developed resistance to some insecticides used in lice shampoos. We first heard about using Listerine against lice in 1999. A reader reported spraying it on her child's head before his possible exposure to lice. A lice expert once told us that she thought the alcohol (26.9 percent) in Listerine was toxic to lice. The herbal oils found in Listerine (thymol, eucalyptol, menthol, and methyl salicylate) may also contribute to the effect.

MAYONNAISE

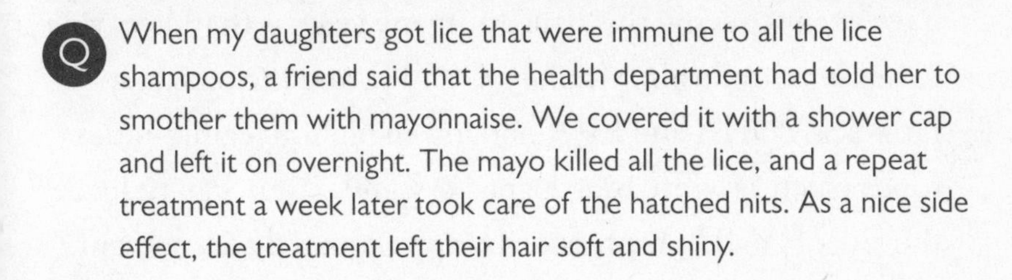

Q When my daughters got lice that were immune to all the lice shampoos, a friend said that the health department had told her to smother them with mayonnaise. We covered it with a shower cap and left it on overnight. The mayo killed all the lice, and a repeat treatment a week later took care of the hatched nits. As a nice side effect, the treatment left their hair soft and shiny.

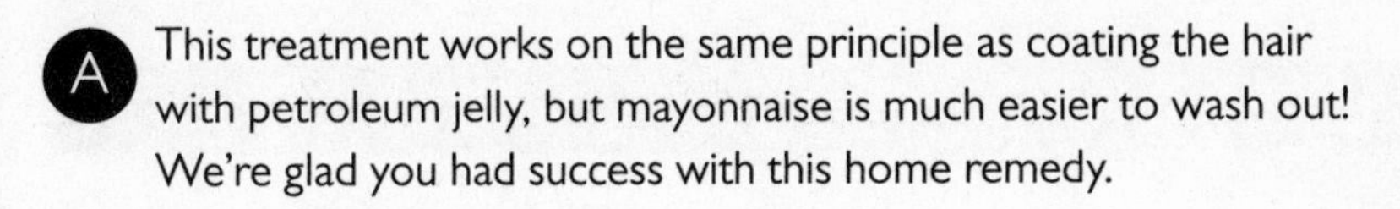

A This treatment works on the same principle as coating the hair with petroleum jelly, but mayonnaise is much easier to wash out! We're glad you had success with this home remedy.

Macular Degeneration

Age-related macular degeneration (AMD) is one of the leading causes of blindness among the elderly in the United States. The condition is caused by a deterioration of the central part of the retina, known as the macula. Because this part of the eye is essential in focusing vision, people with macular degeneration find it difficult to read, to sew, or to do other close work that requires seeing details. AMD is most common among people over 60 years old. (Smoking significantly increases the risk of AMD; any smokers worried about this disease should get help quitting.) Medical scientists believe that heredity plays an important role in determining vulnerability to this condition, but studies also hint at nutritional factors. Recent studies, such as the Blue Mountains Eye Study in Australia, have found that carotene compounds called lutein and zeaxanthin are protective.[1] Lutein and zeaxanthin are found in green and yellow vegetables, such as corn, but they are also abundant in egg yolks. The latest analysis from Tufts University scientists who have been studying this issue for many years is that diets rich in lutein, zeaxanthin, vitamins C and E, and the omega-3 fatty acids DHA and EPA, and low in high-glycemic-index foods (such as white bread, crackers, and pasta), can reduce the risk of AMD progression by approximately 40 percent.[2]

BILBERRY

Q My wife was diagnosed with macular degeneration, and our ophthalmologist said it would just get worse. We immediately started taking bilberry fruit capsules because I wanted to be proactive. A year later, we returned for her annual eye exam. The doctor's assistant administered the exams. After checking my wife three times, she took her folder to the doctor and told him in front of us that the assistant last year sure messed up the exam. The doctor replied, "I administered that exam myself, and I know it is proper." The assistant exclaimed that the macular degeneration was only blocking 25 percent of vision instead of 45 percent like last year, and that was impossible. The doctor asked what we had done, and I told him about taking bilberry. When my wife passed away three years later at age 82, she no longer had macular degeneration.

A Bilberry has a reputation for being good for eyesight. There has been very little research on its power to slow or to reverse macular degeneration in human beings, but there are some intriguing data involving animals.

Q I have recently been diagnosed with macular degeneration. My husband read that the herb bilberry is good for the eyes. I worry, though, that it might interact with any or all of my medicines: Pacerone, Coumadin, and Premarin. Would it be safe for me to try bilberry?

A The idea that bilberry fruit is beneficial to the eyes is based primarily on folklore. However, one study suggests that bilberry can inhibit the overgrowth of blood vessels.[3] Since this is one of the problems that occurs in advanced AMD, it might be a mechanism by which bilberry extract protects the retina from the damage of macular degeneration. Bilberry is not known to interact with medications. Coumadin, however, interacts with many herbs. Ask your doctor to monitor your prothrombin time (a measure of bleeding) more closely if you decide to try this herb in addition to appropriate medical care.

FISH

Q I have been diagnosed with the onset of macular degeneration. The eye doctor said there is no cure. Are there any vitamins or other nutritional supplements that might slow the process down?

A Research has shown that several nutritional factors can slow the development of macular degeneration. One significant study, the Age-Related Eye Disease Study, demonstrated that vitamins C and E, together with beta-carotene and the minerals zinc and copper, could slow vision loss.[4] Additional studies confirmed that these nutrients can help prevent age-related macular degeneration.[5] Research shows that people who eat more fish are less susceptible to AMD, presumably because of the omega-3 fatty acids in fish.[6] Another study found that vitamin D also can reduce the risk of AMD.[7] Research also shows that a diet rich in refined carbohydrates, such as sugar and white flour, is not good for the eyes.[8] People with diets high on the glycemic index were almost 50 percent more likely to develop advanced macular degeneration.

LUTEIN-RICH VEGGIES

Q I've seen lutein listed as an ingredient in Centrum and Ocuvite vitamins. What exactly is lutein?

A Lutein is a yellow plant pigment, part of the carotenoid family that also includes beta-carotene. Scientists believe that this antioxidant is especially important in the retina of the eye and may help prevent macular degeneration, which can lead to blindness. Vegetables rich in lutein include kale, collard greens, spinach, and Swiss chard.

ZINC

Q I have just learned that members of my family have macular degeneration and that it is hereditary. A doctor recommended vitamins with zinc for this condition. How effective are such supplements?

A Ophthalmologists have long recommended antioxidant vitamins with zinc for patients at risk of age-related macular degeneration (AMD). In this condition, the retina deteriorates and central vision needed for reading or driving is gradually lost. One large study demonstrated that for people at risk of AMD, high-dose vitamins with zinc may reduce the likelihood of developing this condition by 25 percent.[9] The supplements in the study provided 500 milligrams of vitamin C, 400 IU of vitamin E, 15 milligrams of beta-carotene, 80 milligrams of zinc, and 2 milligrams of copper.

Motion Sickness, Vertigo, and Dizziness

Feeling motion sick is awful, especially when you're stuck on a boat, plane, or bus for hours and there's nothing you can do to make it stop. The only thing that might be worse is suffering the same symptoms on dry land—then there's not even the prospect of solid ground in your future to take your mind off your distress. There are some conditions, like low blood pressure, that can make a person dizzy, and dizziness is also a side effect of going on or off many medications. If your dizziness persists, you should certainly see your doctor. But if you're just suffering from some garden-variety motion sickness or vertigo, there are a few tried and true remedies that may calm your sad stomach and sense of the wobbles.

ACUPRESSURE

Q Zoloft was prescribed for me after I complained to my gynecologist of feelings of great despair. He recommended Zoloft because he heard positive things about it for menopausal symptoms and believed there were few side effects. Zoloft did take away my feeling of despair. It also obliterated my sense of humor and caused constant forgetfulness. After six years, my husband convinced me to get off Zoloft. So I bought a pill cutter and started to reduce the dose very slowly. My brain retaliated. I became extremely dizzy, to the point of being bedridden. I thought I would not be able to withstand the withdrawal symptoms. Then I remembered having similar vertigo on a cruise ship. Although the Zoloft vertigo was much worse than seasickness, acupressure wristbands worked! I'm now Zoloft free and have discovered that caffeine contributed to my emotional ups and downs.

A We are glad the wristbands helped conquer your dizziness. This side effect can be troublesome when people stop antidepressants like Effexor, Paxil, or Zoloft. Gradual tapering of the dose may help ease other symptoms, such as sweating, nausea, chills, insomnia, and headache.

EPLEY MANEUVERS

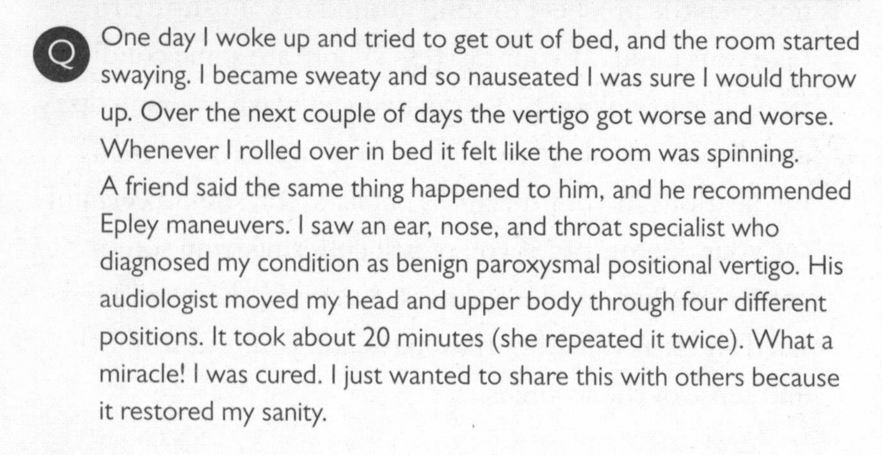

Q One day I woke up and tried to get out of bed, and the room started swaying. I became sweaty and so nauseated I was sure I would throw up. Over the next couple of days the vertigo got worse and worse. Whenever I rolled over in bed it felt like the room was spinning. A friend said the same thing happened to him, and he recommended Epley maneuvers. I saw an ear, nose, and throat specialist who diagnosed my condition as benign paroxysmal positional vertigo. His audiologist moved my head and upper body through four different positions. It took about 20 minutes (she repeated it twice). What a miracle! I was cured. I just wanted to share this with others because it restored my sanity.

A Benign paroxysmal positional vertigo (BPPV), despite its name, does not feel benign to sufferers. Most people who develop this condition are over 60. It is caused by a disturbance in the balance mechanism of the inner ear. Scientists believe this happens when calcium carbonate crystals become dislodged and fall into the semicircular canals of the inner ear. Until recently, there has been controversy over the best treatment. A new review shows that a sequence of head movements, sometimes referred to as the Epley maneuvers, is the most effective treatment. Although people can learn to do the movements themselves, the exercises work best when supervised by a trained professional, such as an audiologist.[1]

GINGER

Q I had vertigo for 17 months, a very severe case. The doctor put me through tests, including an MRI, a test that put me in a black cabinet and spun me around while I answered questions, and one that squirted water into my ear. Then a friend suggested ginger capsules. I took the recommended two capsules four times a day. The vertigo was nearly gone, but I still had a terrible stomachache. I couldn't eat or even sit up. I now take one capsule twice a day. The stomach pain is mostly gone, and so is the vertigo. I can walk up and down stairs without holding onto something, and I can turn around to see something behind me without my head spinning. Will I need to take the ginger for the rest of my life, or will the vertigo stop eventually? In other words, is this a cure or just treating the symptoms?

A Chinese sailors have used ginger for centuries to ease or to prevent symptoms of motion sickness, so we're not surprised it might help treat vertigo. In high doses ginger can cause heartburn or other digestive distress. We cannot speculate about whether ginger will cure your vertigo completely or just relieve the symptoms. At some point you and your physician should evaluate your progress to see whether you can discontinue the ginger.

Q My husband just bought a boat, and I wish I could be as enthusiastic as he is. I get seasick. He wants me to go with him, but I hate feeling so bad. Over-the-counter medicines put me to sleep. Is there a natural remedy for motion sickness that works and won't knock me out?

A You may want to try a preparation containing ginger. One randomized, double-blind study[2] compared ginger capsules with dimenhydrinate, the ingredient in Dramamine. The subjects were 60 passengers on an Italian cruise ship during two days of rough seas. Ginger and dimenhydrinate were equally effective in relieving the symptoms of seasickness, but only 13 percent of people given ginger developed side effects (such as drowsiness and headache) compared to 40 percent of those on the pharmaceutical.

Q After a cruise, I was upset to find that solid ground felt like it was moving. This was very annoying, though it did not make me sick. A friend said ginger worked great for seasickness, so I sliced three pieces of ginger root into hot water and let it steep. The ginger tea made the ground stop moving that same day.

A People often don't anticipate that sensation of solid ground swaying beneath their feet after they have become accustomed to being on a boat. We're glad to hear that ginger tea worked as well for that strange feeling as it does for actual seasickness.

Muscle and Leg Cramps

Having a muscle suddenly and violently contract, and refuse to relax, is a painful experience. It is bad enough when it happens during the day (often after exercise). But when a muscle cramps up at night, it can wake you from a sound sleep. Doctors once prescribed quinine, but the U.S. Food and Drug Administration no longer permits quinine prescriptions for any condition other than malaria because of the dangerous effects of this drug. Those who know they do fine with tonic water may be able to get quinine that way. Others may benefit from some of these home remedies.

AVOIDING EARL GREY TEA

Q You briefly mentioned Earl Grey tea causing leg cramps and suggested that the oil of bergamot used for flavoring might interfere with potassium absorption. Eliminating the tea was an instant cure for me.

A Earl Grey tea gets its distinctive flavor from the citrus fruit bergamot. The oil contains a natural compound called bergapten, which can interfere with the flow of potassium into and out of cells. According to one study, this is presumed to be the reason that too much Earl Grey tea can cause muscle cramps in susceptible people.[1] We're glad you were able to conquer your muscle cramps by giving up the tea.

BLACKSTRAP MOLASSES

Q Years ago, I heard that a tablespoon of blackstrap molasses would alleviate leg cramps. I used to rely on it but haven't found it recently.

A Ask your grocer or check a health food store. Blackstrap molasses is rich in potassium and iron and has some calcium and magnesium.

CASTOR OIL

Q For about 20 years, I had chronic pain from a muscle knotted up in my back, like a constant cramp. I tried unsuccessfully to address it with chiropractic adjustments. A massage therapist recommended applying castor oil, covering the area with felt, and then putting a heating pad on the sore spot. It has taken three treatments, but the results are dramatic.

A Castor oil applied topically has been reported to ease bruising. We're impressed that it relieved your muscle pain.

GATORADE

Q I cramp easily even when I have not done anything strenuous, but more so after playing tennis. Drinking Gatorade before and after tennis helps some. What am I lacking?

A If Gatorade helps somewhat, you may be low in some electrolytes. We have heard from another tennis player that drinking Pedialyte after a match can help prevent cramps. This liquid formulation is designed to help replenish lost fluids and minerals for babies who become dehydrated from diarrhea or vomiting.

GLUTEN-FREE DIET

Q I thought my serious leg cramps were just old age creeping up on me. I handled them with additional calcium and magnesium supplements, and also with tonic water and mustard. It was only when I was diagnosed with celiac disease (I was so anemic that I had to go to the ER) that I realized I was not absorbing calcium. Since I started the gluten-free diet for celiac, I haven't had any leg cramps at all.

A People with celiac disease must avoid gluten found in wheat, rye, and barley. It triggers a reaction that harms the lining of the small intestine and interferes with the absorption of crucial nutrients, including calcium and other minerals.

MAGNESIUM

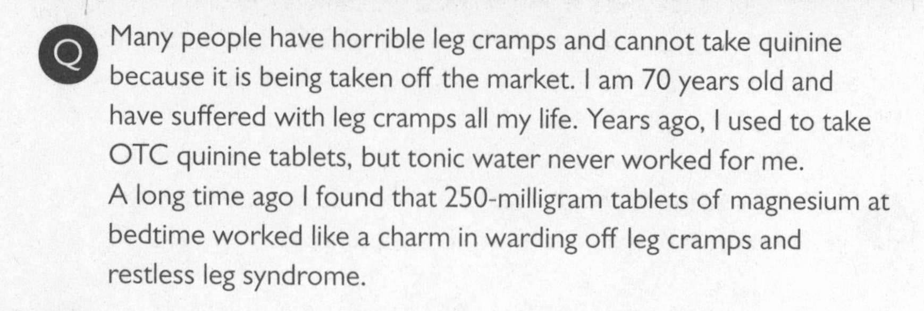

Q Many people have horrible leg cramps and cannot take quinine because it is being taken off the market. I am 70 years old and have suffered with leg cramps all my life. Years ago, I used to take OTC quinine tablets, but tonic water never worked for me. A long time ago I found that 250-milligram tablets of magnesium at bedtime worked like a charm in warding off leg cramps and restless leg syndrome.

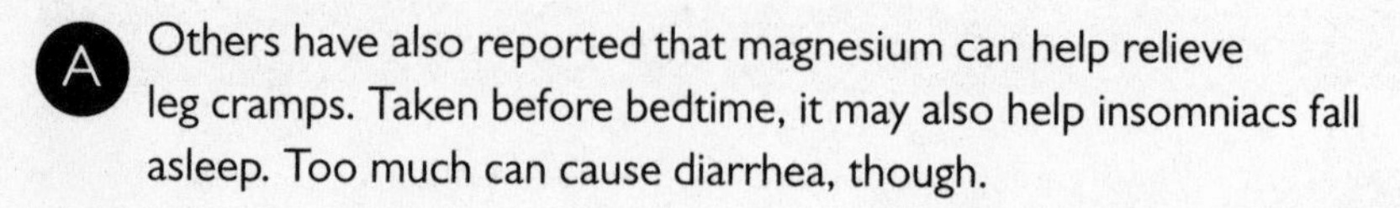

A Others have also reported that magnesium can help relieve leg cramps. Taken before bedtime, it may also help insomniacs fall asleep. Too much can cause diarrhea, though.

MINERALS

Q When I was teaching elementary school, I spent many sleep-deprived nights because I was awakened with leg cramps so severe they took my breath away. I mentioned this to my mother, and she said, "Take calcium-magnesium." I was already taking calcium, but her doctor had told her it must be the combination supplement. I took this with a grain of salt, thinking, "I'll try it; it won't work." As you might have guessed, it did work. I continue to take calcium-magnesium each night and no longer have cramps. One exception: I bowled six games in succession and later that afternoon awoke from a nap with an excruciating cramp in my leg. I took three calcium-magnesium tablets with a glass of water, and the cramps were soon gone. Others might like to know about this.

A Thanks for the suggestion. Calcium, magnesium, and potassium are all essential minerals for muscle function. Other readers have also had success using magnesium or calcium supplements to prevent leg cramps. Taking them together is a good idea.

PICKLE JUICE

Q I'm 47 years old and play basketball. My calves seem to cramp almost every time I play. I stretch them for about 20 minutes beforehand, but it doesn't help. Would this be due to a lack of calcium or potassium? Please help. I'm sick of hobbling around for days after I play.

A Muscle cramps can be caused by all sorts of things. Lack of minerals like calcium, magnesium, and potassium may be responsible. In such cases, replenishing the minerals may prevent the cramps. A former football player with the Dallas Cowboys and San Francisco 49ers told us that the best home remedy is a jigger of pickle juice. You might also want to do your calf stretches after you play, in addition to your exercises before the game.

Q My husband had severe leg cramps for years. While he was at the eye doctor, the receptionist excused herself, saying, "I've got leg cramps. I've gotta grab the dill pickle juice!" When my husband got leg cramps a few nights later, he grabbed the dill pickle jar and poured himself a swig of juice. Almost instantly, the cramps were gone! Once we were out of pickle juice, and he took a tablespoon of mustard. VOILA! He got the same result! Now he keeps little packets of mustard in the car and the truck just in case.

A Thanks for sharing the remedy. Leg cramp victims may well benefit.

SOAP

Q I read in your column about putting soap under the sheet to stop leg cramps. My husband tried it, and it worked. Before, his legs had sometimes hurt so bad he would almost cry. Once he tried the soap, his legs hurt only after he got out of bed. Now he keeps a motel-size bar of soap with him at all times and has no more leg pain during the day either. Our friends think we have lost our marbles, but who cares? No leg pain!

A Many doctors think we have lost our marbles for recommending this strange remedy. Nevertheless, we have heard from many readers that it helps, and we can't see how it would hurt.

Q I read your column about using soap in the bed for leg cramps. It really works. My husband had leg cramps for years, since he is a fisherman and on his feet 18 hours a day. They brought tears to his eyes, although he is very stoic. He's had no more leg cramps since we started putting soap under the bottom sheet. We were playing cards the other night, and he got cramps in his hands, holding the cards, as he sometimes does. I got a bar of soap and put it in his hand. Within a minute the pain subsided. He held the bar for about ten minutes, and the cramp never came back. Now we keep a bar of soap near where we play cards.

A Though many people have told us of their success using soap under the bottom sheet to ward off nighttime leg cramps, this is the first we have heard of using soap to keep away hand cramps.

TONIC WATER

Q While I was recovering from surgery, I had to wear a cast on my leg, and I got leg cramps. A friend recommended tonic water. BINGO! It worked like a charm.

A Thanks for sharing your success with tonic water. Many brands of tonic water contain quinine, and other readers have also reported finding it helpful against leg cramps. Quinine has been used medicinally to ward off malaria since the 17th century. A glass of tonic has roughly 20 milligrams of quinine, which is a relatively small dose. Nevertheless, some people are so susceptible to serious side effects from quinine that they must avoid even this small amount. For them, quinine may cause life-threatening heart rhythm disturbances, severe skin reactions, and several blood-related complications. That is why the FDA banned quinine for treating leg cramps.

TURMERIC

Q Mustard works for leg cramps. But I really don't like it straight up, especially at night before going to bed. Turmeric is the active ingredient in mustard. When I get leg cramps, I mix one-quarter teaspoon turmeric in four ounces of water and drink it down. Leg cramps subside in a minute or less, faster than mustard and a whole lot more palatable.

A **Thanks for the tip. Some may find the taste of pure turmeric even more challenging than mustard, though. Turmeric is more soluble, and may be more palatable, in milk.**

V8 JUICE

Q A year ago your column mentioned the benefits of low-sodium V8 juice for muscle cramps. For years I had suffered from severe leg cramps almost nightly. I would awaken in agony, even though I was eating bananas and taking potassium supplements daily. After seeing your column, I began drinking eight ounces of low-sodium V8 juice every day. Now, more than a year later, I have not had one episode of muscle cramping. V8 juice has fewer calories than bananas, and it's the solution to a painful problem.

A **Several readers mentioned the high potassium content of low-sodium V8 juice (840 milligrams in 8 ounces). This offers more potassium for fewer calories than either bananas or orange juice. We're glad it has prevented your leg cramps.**

Q I awoke one night with the muscles and nerves in my legs feeling like fireworks (little explosions . . . big explosions . . . the grand finale!). They were so active that they started twisting into cramps. This literally kept me awake all night. The cramps were in all parts of my legs and feet—nowhere else in my body. Blood tests didn't reveal any problems, so the doctor shrugged and gave me muscle relaxants so I could sleep at night. I took them for months. If I stopped the drug, the problem returned. One day I read in your column about a man who was taking a diuretic and began having cramps in his legs at night. He found that low-sodium V8 stopped the problem. Since I was taking a diuretic at the time, I thought I would give it a try. It stopped the cramping and severely curtailed most of the weird muscle and nerve activity. I faithfully drink two glasses of the V8 every evening, and I haven't taken a muscle relaxer since the first glass.

A Low-sodium V8 juice provides plenty of potassium. When this mineral is in short supply, many people develop cramps. Diuretics frequently deplete the body of potassium, and that may be why low-sodium V8 helps some people. Anyone who is taking angiotensin-converting enzyme (ACE) inhibitor blood pressure medicines like enalapril, lisinopril, or ramipril, though, must be very cautious about extra potassium.

VINEGAR

Q I sometimes get painful leg cramps that wake me up in the middle of the night. The best treatment I have found is to drink vinegar. I put one tablespoon in six to eight ounces of water with one-half teaspoon of sugar to help it go down. The leg cramps are gone in about five minutes. I carry a small bottle of vinegar with me now just in case.

A This sounds like a very easy solution. Thanks so much for telling us about it.

YELLOW MUSTARD

Q We tried a treatment from your column for nighttime leg cramps. My husband used to get them frequently and would have to walk them off while in pain. He read that taking mustard would alleviate them, so he tried it. Now when he gets leg cramps at night, he takes his mustard and they go away quickly. He keeps a few individual packets of mustard in the bedroom. He thought it was just an old wives' tale, but now he's a believer.

A We are delighted to learn that yellow mustard has helped relieve your husband's leg cramps. A retired pharmacist told us about this remedy nearly six years ago: "A friend of ours uses plain mustard for leg cramps. She swallows a teaspoonful of mustard to relieve the pain. This home remedy works so well for her that she carries packets of mustard wherever she goes." Since then we have heard from many folks who use yellow mustard to relieve leg cramps. Although there is no scientific proof, we suspect that turmeric, which gives mustard its yellow color, may have a beneficial effect.

Q I suffer from leg cramps. Recently, while attending a basketball game, I had to leave my seat and try to walk off a severe inner thigh cramp. A security guard, seeing that I was grimacing in pain, approached me to see if I needed first aid. When I said it was only leg cramps, he took me to the concession stand and suggested I try yellow mustard. I asked if I was supposed to eat it or apply it. He said it was an old-time remedy his grandmother used. I ate the mustard. By the time I walked to the end of the concession stand, my leg cramp was gone! I have since used this remedy repeatedly, especially in the middle of the night when my cramps seem to occur most often. It works. I don't know why, but I sure am glad it does.

A You are not the first to share the secret of yellow mustard. Some readers keep little packets of this condiment on their nightstand to ease leg cramps, although others have complained that mustard can give them indigestion.

Nail Fungus

For most people, nail fungus is a minor condition that requires no treatment. Thick yellow or brown nails aren't pretty, and they can be hard to trim properly. That is the primary reason this infection is more serious for people with diabetes, who are prone to foot problems. But for the rest of

THE PEOPLE'S PHARMACY

Favorite Food #21: Mustard

Mustard dates to the Romans, who apparently combined unfermented grape juice (*must*, or "new wine," in Latin) with ground mustard seed. At first people prized mustard more for its medicinal properties than for its tingle to the taste buds. Hippocrates used mustard in poultices and various medicines. By the 1500s, Europeans frequently used mustard plasters to ease chest congestion. Up until the early 20th century, physicians in the United States routinely recommended mustard plasters on the chest for treating lung congestion and a cough.

We enjoy all sorts of mustards, from old-fashioned French's yellow hot dog mustard to German mustard with the seeds, dark deli-style brown mustard, and Dijon mustard. Mustard's basic ingredients include mustard powder, vinegar, water, salt, sugar, and various spices. In the United States, mustard is usually made with white mustard seeds. To make mustard the yellow color we expect at the ball park, the herb turmeric is added.

We weren't particularly aware of any health benefits linked to this spice until our readers started sharing some amazing stories with us. One long-distance bicyclist related his story of using mustard to relieve leg cramps.

On a ride of over 100 miles, his legs started to cramp around mile 75. He had heard that plain yellow mustard could help calm leg cramps, and he had brought a couple mustard packets (the kind you get at a fast-food joint) with him. He "sucked them down," and, to his amazement, the cramps disappeared almost immediately. He was able to finish the ride without more problems.

Having heard from so many people about the unique benefits of mustard, we did some research. We learned that mustard seeds contain omega-3 fatty acids (the good fat in fish oil), as well as magnesium, iron, calcium, selenium, and several other important minerals. One study revealed that mustard oil was more effective than fish oil in preventing colon cancer.[1]

us, home remedies might be just as rational as prescription medications such as Lamisil or Penlac. Although approved by the U.S. Federal Drug Administration, these pricey drugs don't work for everyone. Oral drugs such as Lamisil may have systemic side effects. Home remedies may be no more effective, but at least they are inexpensive.

CORNMEAL

Q I tried the suggestion of using cornmeal mush for toenail fungus. I used it twice, and after the nail grew out the fungus was gone. My problem was only with the big toenail, not all the toes.

A We heard about using cornmeal to fight nail fungus from a gardener in Vicksburg, Mississippi. He told us to put an inch of cornmeal in a footbath and cover it with hot water so it can dissolve. Let it cool so you don't burn your feet, and then soak them in this mush for an hour. This is an inexpensive and low-risk approach to nail fungus.

DANDRUFF SHAMPOO

Q My neighbor swears that brushing his toenails with Pert Plus or Head & Shoulders dandruff shampoo when he showers both cures and prevents toenail fungus. He claims that if you wash your hair with these shampoos, you will never have fingernail fungus. Is he nuts, or is he on to something?

A Dandruff is caused by yeast (a type of fungus). That's why dandruff shampoos with antifungal ingredients are effective at controlling flaking and itching. It is conceivable that brushing toenails with dandruff shampoo might provide some protection against fungal infection. Once a nail is infected, though, it is unlikely that dandruff shampoo will solve the problem. Toenails are tough, and it is difficult for antifungal products to penetrate them.

HYDROGEN PEROXIDE

Q I have battled toenail fungus off and on for the past 25 years. I have been on Lamisil three times and tried all sorts of OTC and prescription topical medicines. I decided to try two of the remedies I read about in your articles. I applied hydrogen peroxide with a cotton ball to my toenails after I bathed daily. Then I applied VapoRub to my feet and toenails and put on socks to sleep in. Within a month, I had no more toenail fungus. I have the most beautiful toenails I have ever had in 25 years. I also like the fact that I can polish my toenails and still use these remedies. Thank you!

A **Toenail fungus can be tough to treat. Prescription medicines like Lamisil are expensive and require medical monitoring for liver problems and other potential side effects. Success with home remedies like the ones you are using requires persistence. Hydrogen peroxide is antimicrobial, and Vicks VapoRub contains herbal oils with antifungal activity.**

Q I tried your hydrogen peroxide treatment for toenail fungus, and it worked like a charm.

A **The reader who suggested this remedy applied pharmacy-strength hydrogen peroxide daily to nails with a cotton ball after showering.**

IODINE

Q For over 20 years I was plagued with dry, flaky skin on the side of my nose and behind my earlobes. I went to several doctors, including dermatologists. We tried various salves to no avail. After a while this added up to big bucks. I wondered if this ailment was caused by a fungus. As a chemist, I know that iodine is very effective on fungus and many bacteria. I applied tincture of iodine with my fingers (every two days for a week) and got cured within a week. It's been two months, and the spots have not returned. The cost of treatment was less than two dollars. I previously had success treating toenail fungus with iodine. Since I am a man, dark toenails (stained by iodine) did not bother me. I have suggested iodine treatment to many friends, and the responses have been enthusiastic. Many have solved their nail problems after other pricey treatments were unsuccessful.

Iodine was discovered about 200 years ago. As a tincture, it has broad antiseptic activity. You are correct that it kills fungus. Tincture of iodine is dark brown and can stain, so not everyone will want to use it on the face. An alternative would be clear, decolorized or "white" iodine (*yodo blanco* in Spanish). It won't stain the skin, and many readers tell us it works as well as tincture of iodine for fighting fungus-infected nails.

LISTERINE

When I needed treatment for toenail fungus, my doctor suggested I soak my toes in Listerine for 30 minutes a night for 30 days. I sent my husband to Costco for a giant jug of Listerine. He returned with the minty one. It's blue, but I figured that wouldn't really make a difference. It did. My feet turned blue, and no amount of scrubbing could take the color off. My husband laughed until he cried. After switching to the regular (amber) Listerine, my toenail fungus did clear up, but the nails themselves were dry.

The herbal oils found in old-fashioned Listerine, such as thymol and eucalyptol, have antifungal properties. Many of our readers have found that Listerine is effective in fighting nail fungus. The alcohol in regular Listerine (26.9 percent) might be the culprit that dries out your nails.

LISTERINE AND VINEGAR

Q You receive many letters about nail fungus, and I wanted to share my experience. Our daughter contracted a foot fungus while swimming at a local club when she was six. We've tried a lot of different antifungal products, but I didn't want to give her oral medicine. The podiatrist suggested a mixture of half white vinegar and half Listerine. I dabbed it onto her toes every morning with a cotton ball. Finally her toenails are pink and healthy looking. It works, but it takes a very long time.

A We first wrote about using a mixture of white vinegar and Listerine for nail fungus in spring 2005 after hearing about its potential from one reader. Some people dab it on their nails, while others soak their feet in the solution. (It can be reused several times.) The herbal oils in Listerine have antifungal properties, as does the alcohol. Vinegar is acidic, and thus discourages the spread of fungus. Perhaps the combination of Listerine and vinegar provides more antifungal activity together than individually.

Q I've had toenail fungus on all my toes for a number of years and would like to know the most effective treatment.

A There are no studies to show effectiveness of most home remedies, so it is hard to predict what will work best. Whether you treat your nail fungus with a prescription drug or a home remedy, you'll need to be persistent. Here is one reader's experience: "I read your column about toenail fungus and athlete's foot. I've found the foolproof method to get rid of this problem without using toxic prescription meds. Soak the feet twice a week for about five to ten minutes in a 50/50 mixture of white vinegar and regular (amber) Listerine. The mixture can be put in a sealable container large enough for your feet and can be used for three months before making another batch. For in-between use and traveling, pick up a Misto container for spraying cooking oil at any good kitchen store. It's unbreakable, doesn't leak, and travels well. I use it to spray my feet daily, especially between the toes. It takes about a year of this to clear up a serious toenail fungus, but it also eliminates other foot problems caused by moisture."

TEA TREE OIL

Q I developed toenail fungus on both big toes but was reluctant to use harsh pharmaceutical treatments. So I blended a one-to-one ratio of tea tree oil (antifungal) and DMSO (penetrating carrier) and added a few drops of clove oil (which will kill anything—fungal or bacterial!). I shook it up well and dabbed it on my nails twice a day, making sure they were dry before I put on my shoes. It took three weeks, but my nails are now clear. I love home remedies, so I thought I'd pass this one on!

A Dimethyl sulfoxide (DMSO) is an interesting compound. This solvent carries other chemicals through the skin, and is used in some topical prescriptions products for this purpose. Tea tree oil has demonstrated antifungal properties, and the DMSO might help it get through the nail. Your formula is unique, but it may be irritating to the skin. Anyone who tries it should restrict this mixture to the nails.

URINE

Q Over the last nine years I have tried everything I ever heard about and everything you have written on treating nail fungus. Nothing worked. Then I read your column about the soldier who was told by his sergeant to urinate on his feet while in the shower. Well, I was desperate, so I urinated in a foam cup and then soaked my finger for about five minutes every night. After soaking, I cleaned my finger with a hand sanitizer. I know it sounds gross. I did this for about three weeks. That was three years ago, and the fungus is still gone!

A A few years ago we heard from both a grandmother and a World War II veteran that urinating on your feet in the shower could help control athlete's foot. This is the first time we have learned that this remedy might also work against nail fungus. Thanks for sharing your experience with this inexpensive remedy. We cannot explain how it might work.

VINEGAR

Q My sister has been going through chemotherapy and radiation for breast cancer this past year. She started losing her toenails due to fungus and was shocked at the cost of medication for this. I passed along your article on vinegar soaks. She laughed at first but decided to try it. Guess what? She hasn't lost another nail, and the fungus has gone away. She even copied the article and gave it out at the chemo center! Thanks.

A We are pleased your sister has benefited and shared the word. Vinegar soaks are a simple, inexpensive approach to nail fungus. The acid in vinegar makes for an inhospitable environment for fungus.

Q My husband's toenail fungus was terrible. I read your article about using vinegar and water to heal the fungus. My husband started this treatment a few months ago and soaked his feet almost every night. The toenails on one foot are completely healed, and the other foot has only one nail still affected. Thanks for such good advice.

A Soaking the feet nightly in a solution of one part vinegar to two parts water is a remedy that seems to help many fight off nail fungus. It takes patience, since the nail has to grow out fungus free. That may take many months. If vinegar soaks do not help, there are several other home remedies that can be helpful, such as applications of tea tree oil, vitamin E oil, hydrogen peroxide, rubbing alcohol, or iodine.

Nausea

The queasy feeling that precedes vomiting has a number of causes. It may be the first alert of a gastrointestinal infection or a reaction to medication. The nausea that accompanies pregnancy, especially in the first trimester, is notorious, and pregnant women are cautioned against taking medications. So it is helpful to have a couple home remedies to try.

ACUPRESSURE

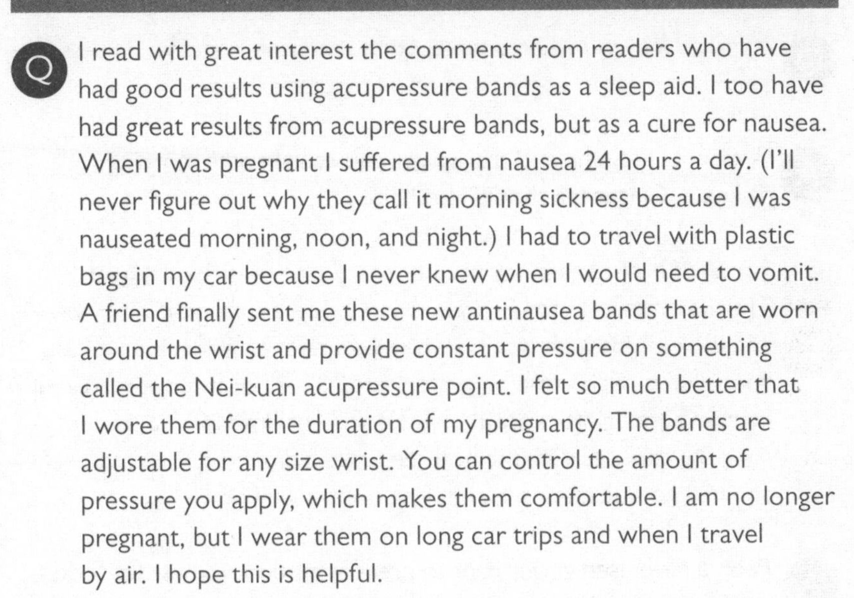

Q I read with great interest the comments from readers who have had good results using acupressure bands as a sleep aid. I too have had great results from acupressure bands, but as a cure for nausea. When I was pregnant I suffered from nausea 24 hours a day. (I'll never figure out why they call it morning sickness because I was nauseated morning, noon, and night.) I had to travel with plastic bags in my car because I never knew when I would need to vomit. A friend finally sent me these new antinausea bands that are worn around the wrist and provide constant pressure on something called the Nei-kuan acupressure point. I felt so much better that I wore them for the duration of my pregnancy. The bands are adjustable for any size wrist. You can control the amount of pressure you apply, which makes them comfortable. I am no longer pregnant, but I wear them on long car trips and when I travel by air. I hope this is helpful.

A Thank you for sharing your story. Others have also reported success with wrist acupressure for nausea or insomnia. Those who are curious can search the Web for Psi Bands, Sea-Bands, or AcuBands.

Q Is there any nondrug solution to extreme nausea during pregnancy? I am four months pregnant and nauseated all the time. Nothing my doctor has suggested has worked more than a week. I am desperate.

A Morning sickness is a misnomer for many women, since the nausea and vomiting can occur anytime. For some, the condition is so severe that they may lose weight and become dehydrated, putting the baby at risk. One device that may help is called PrimaBella. It looks a lot like a wristwatch, but it delivers a mild pulse of electricity to an acupressure point on the inner wrist. This electronic tool is available by prescription for morning sickness, chemotherapy, and postoperative nausea, and without a prescription for motion sickness. You can learn more from the website: *www.primabellarx.com.* A meta-analysis has also indicated that stimulation of the P6 acupuncture point was about as effective as antiemetic drugs for treating postoperative nausea.[1]

GINGER

Q It makes me mad when people won't try home remedies. I guess it sounds too easy. Ginger tea works great for nausea.

A A recent study showed that ginger pills can alleviate morning sickness. Ginger tea should also help.

Q For all of his 65 years, my partner has gotten seasick on a boat. This was true even for two years serving on a ship in the navy. We just took an Alaska cruise on which I served as a guide. He felt just fine, even though some others did not. Why? Because we read your column about ginger capsules, available in health food stores. He took the capsules with meals three times a day. He got great benefit and had no side effects. I am telling everyone I know!

A People have used ginger root to prevent motion sickness for thousands of years. Placebo-controlled studies confirm its effectiveness.

Q Can you tell me about ginger for nausea, especially during pregnancy?

A People in China have used ginger root to combat nausea for centuries. Several studies suggest it may also fight morning sickness. One reader reports, "I taught childbirth preparation classes for a number of years and suggested candied ginger to my students. It has a long shelf life, especially when refrigerated. Candied ginger is inexpensive, readily available, and often helpful."

Nerve Pain

If you hit your thumb with a hammer, the cause of the pain is clear. Not so with the mysterious ailment known as neuropathy, or nerve pain. We know that chronic conditions such as diabetes can cause peripheral neuropathy, or nerve pain in the legs (and very occasionally in the hands and arms). Some medications provoke neuropathy as a side effect; statins may be among the most commonly prescribed drugs that produce this effect. Shingles is itself a very painful affliction, but for some people the nerve pain that lingers after the rash has healed (known as postherpetic neuralgia) is lasting and excruciating. According to one study, vitamin deficiencies, particularly in vitamin B_{12}, can lead to nerve pain.[1] In that case, the pain should respond to vitamin supplementation, either oral or injected.[2] But it is by no means obvious why some people develop trigeminal neuralgia, which is pain that runs along the path of the trigeminal nerve in the face and head.

Doctors have a few medications that they often prescribe for pain rooted in irritation of the nerves. Medications initially developed for treating epilepsy are often helpful: Neurontin, Topamax, or Lyrica. In some cases physicians prescribe antidepressants for the pain itself rather than for the low mood it might understandably produce. Old medications like amitriptyline or nortriptyline, and newer ones like Cymbalta, may be used in this manner.

Research shows that diabetics with peripheral neuropathy may get relief from alpha-lipoic acid, a nonprescription dietary supplement.[3] Many readers are enthusiastic about the value of the herb turmeric, either alone or in combination with bromelain and boswellia, but at this time the research is limited to investigating basic mechanisms of its action in animals.[4]

TURMERIC

Q: Turmeric is excellent for nerve damage. I had a pinched nerve in my neck. I went to the emergency department, but the doctors said they

THE PEOPLE'S PHARMACY

Favorite Food #22: Ginger

Ginger comes from the rhizome (or underground stem that is commonly referred to as a root) of the Southeast Asian plant *Zingiber officinale*. The flavor is spicy, almost hot, though in a completely different way from hot peppers. In addition to proving useful as a flavor in spice cookies or curries, ginger has been widely used to calm the digestive tract, to ease symptoms of colds, and to provide pain relief.

Chinese sailors used to rely on ginger to prevent motion sickness, and some research suggests they may have been on to something.[1] Although some investigators have not found ginger to be effective against motion sickness,[2] most of the evidence indicates that it can be helpful to combat morning sickness.[3]

Scientists are also investigating ginger for its ability to quell the debilitating nausea associated with chemotherapy.[4] In a study of more than 600 patients undergoing chemotherapy, participants were given a standard antivomiting medication. They also took either ginger or placebo capsules for six days, starting three days before chemotherapy.

They rated the severity of nausea four times daily during treatment. The patients who took ginger had 45 percent less nausea than they had experienced in a previous round of chemotherapy with only standard medication. Those who took the placebo did not have a significant difference in their ratings.

Other readers have found that ginger has additional uses as well. It can ease the symptoms of a cold or cough, restore a hoarse voice, or help alleviate a headache. Though little research demonstrates that ginger is truly effective for colds and headaches, one reader reported on ginger's effectiveness for a persistent cough: "I tried antibiotics, gargles, and anesthetic lozenges, but none of them worked. Then my father suggested that I chop up some raw ginger root, chew the pieces like candy, and suck the juice out of them. I tried his suggestion, and the following day, my cough was completely gone."

had no painkiller for nerves. Instead they gave me a high-powered pain pill. Those pain pills almost killed me—my reaction to them set the nerve on fire. The pain was so great that I cried all the time. For 14 days I couldn't sleep for more than a few minutes here and there. Once I stopped the pain pills, the pain eased up somewhat. I asked at the health food store if there was anything to help the pain in my neck. They recommended turmeric, so I got a bottle, and with the very first pill I could feel the pain going away. The turmeric I take is in a combination formula. I use Solaray Turmeric Special Formula. It combines turmeric with bromelain and boswellia, and I have gotten good relief from it for five years.

Q I've found turmeric to be very helpful for pain associated with multiple sclerosis. I gave up on all the medications prescribed for MS pain—none gave me any relief. But I've taken two capsules of turmeric daily for almost two years and have had dramatic relief from pain in my feet and legs.

A **Bromelain, boswellia, and turmeric are all traditional Indian ayurvedic medicines with strong anti-inflammatory properties. Bromelain is an enzyme derived from pineapple. Boswellia comes from a tree resin similar to frankincense. Turmeric is the yellow spice in curry powder and yellow mustard. Other readers have also reported that this combination is especially helpful for nerve pain.**

VITAMIN B_{12}

Q My fingers and toes started tingling unpleasantly about three months ago. My internist recently did blood work and found no diabetes or other problems. When the nurse asked me if I was still taking my usual medicines and vitamins, she mentioned B_{12}. It dawned on me then that I had discontinued it when I was suffering from a nasty coughing spell. The tingling began about the time I stopped taking the vitamin B_{12}. The doctor wanted to refer me to a neurologist, but I asked if I could delay that until I began taking B_{12} again to see if the tingling disappeared. I've just begun taking the supplement, but it seems to be working. Others might like to know about this.

A Tingling or numbness in the hands or feet is called peripheral neuropathy. A fairly common complication of diabetes, it can be caused by some medications, including certain cholesterol-lowering drugs, cancer treatments, and antimicrobials. Inadequate vitamin B_{12} can also contribute to neuropathy. We hope that returning to your regular vitamin regimen will relieve the tingling.

Nosebleed

If you get hit hard in the nose, there's a good chance you will end up with a nosebleed. No mystery there. Most nosebleeds occur for no apparent reason, though. Children are especially susceptible. Chronic or severe nosebleeds require medical attention to rule out any serious underlying medical condition. There are effective over-the-counter medications. If, on the other hand, a nosebleed happens at work or at night when it is inconvenient to get to a pharmacy, a home remedy may be just the ticket.

COLD KEYS

Q My daughter has bad nosebleeds. Do you have any herbal or home remedy suggestions?

A You may want to start in the pharmacy. There are three products to consider: Nosebleed QR (*www.biolife.com;* 800-722-7559), NasalCEASE (*www.nasalcease.com;* 800-650-6673), and Seal-On (*www.seal-on.com*). As for home remedies, our favorite would be to put a ring of cold keys down the back of her neck under her shirt. We cannot explain why this might work, but we have heard from many readers that it is amazingly effective. One reader offered this: "When I was a little girl in rural North Carolina, my daddy knew how to stop nosebleeds when someone in the family had one. He put a bunch of car keys down our backs. The nosebleed stopped pronto."

Q When I was a kid, I got very bad nosebleeds. If nothing else worked, my mother would get out her keys and drop them down the back of my neck. I wish I knew why it worked so well.

A We have heard from many people who have had success stopping nosebleeds with keys or a cold butter knife against the back of the neck. We don't know why this trick works, but one reader offered the following from his experience as a medic doing water rescue: "The keys work because of the mammalian dive reflex. Cold hits the nerves in the neck, causing the blood vessels to constrict. You might notice your pulse slowing too. The dive reflex is why cold-water drowning victims are not usually pronounced dead until they are 'warm and dead.' Cold water only in the face and head area shunts blood to the organs and away from the skin and slows the metabolism for survival. The vital signs are often too weak to detect." This hypothesis sounds plausible to us. We can't offer a better one.

PAPER

Q I had nosebleeds from infancy to late puberty. An uncle gave my mother this trick. He served in the army in WWI, so it goes way back. Take some brown paper from a shopping bag, fold it into a double strip about an inch and a half long by one-quarter inch, and place it in the front of the mouth as far up as you can, so it applies pressure above the jawbone and just under the nose. It will stop a torrent in a minute or so. It is quicker than cold keys on the back of the neck.

A We've heard of this remedy, but we cannot explain how it would work, any more than we can explain how keys dropped down the back of the neck would stop a nosebleed.

Plantar Fasciitis

This heel condition hurts like hell. Typically, you feel the pain on the underside of the heel through the arch, and it is especially uncomfortable when you first get out of bed in the morning and take a step or two. Wear and tear leads to inflammation of the plantar fascia, which supports the arch. A podiatrist is likely to recommend special stretching exercises, orthotics, good shoes, and rest. Home remedies may also aid in recovery.

CHERRY JUICE

Q I was diagnosed with plantar fasciitis by one of the best foot doctors in my city. I was given pain medicines, many anti-inflammatory drugs, and foot splints with no success. As a last resort, he recommended steroid injections for the intense pain. A friend suggested that I try cherry juice. In two days, I was nearly pain free. It was almost a religious experience. I am convinced this works, and I have since drunk more cherry juice when pain flared up a few weeks later. Once again I got great relief.

A A number of animal studies on rats treated to develop arthritis have shown that cherry extract can reduce paw swelling and pain behaviors. The red compounds, anthocyanins, appear to have anti-inflammatory effects. We don't know why cherry juice would have worked when anti-inflammatory drugs such as aspirin did not, but we're glad to hear of it.

Q I have a sensitive stomach, so drugs like ibuprofen and naproxen cause me problems. I have also heard that these drugs can be hard on the liver and kidneys. What else can I use for my plantar fasciitis and back pain?

A Nonsteroidal anti-inflammatory drugs (NSAIDs) can certainly cause stomach ulcers and kidney or liver damage when taken in high doses for long periods of time. Some people can't even tolerate low doses without getting heartburn. For plantar fasciitis, the best solutions are arch supports and stretching. Muscle pain may also respond to tart cherry juice. Two separate research projects on horses and college students have shown that cherry juice minimizes exercise-induced muscle damage.[1]

FISH OIL

Q I am an RN and am constantly on my feet at the hospital. I was diagnosed with plantar fasciitis two years ago and had steroid injections in both heels several times without much relief. A couple weeks prior to seeing the podiatrist who would do surgery on my heels, I began taking fish oil for my cholesterol. I was a little embarrassed when I went for the presurgical visit because at that time I wasn't having any discomfort. I told him about the fish oil I was taking, and he told me that fish oil has anti-inflammatory properties. I take four capsules a day, so I don't need the surgery and have no pain!

A Your podiatrist is quite right. Research indicates that the omega-3 fatty acids in fish oil have anti-inflammatory activity,[2] and taking roughly 4,000 milligrams per day should certainly be enough to make a noticeable difference. Other conditions that may respond favorably to fish oil include attention deficit disorder (ADD), osteoarthritis, heart disease, depression, and dementia.

Psoriasis

Psoriasis is an autoimmune condition. The condition is mild but chronic, and there is no true cure. In psoriasis, skin cells rev up and accelerate their life cycle. This results in reddish patches, often with silvery scales, on the elbows, knees, back, or scalp. People with psoriasis may feel embarrassed because it can be scary looking; others may need to be reassured that it is not contagious in any way. Sometimes psoriasis affects toenails or fingernails. At times psoriatic nails may be mistaken for nails infected with fungus, but antifungal medicines will be of no help. Psoriasis is an inflammatory disease, and it can also affect joints. Psoriatic arthritis can be quite debilitating and may require heavy-duty medications such as methotrexate or newer ones like Remicade or Enbrel, which are also used for rheumatoid arthritis. In some cases, sun exposure (but not enough to burn the skin) can help tame the plaques of psoriasis. In others, a doctor will prescribe oral medicine to sensitize the skin before the patient is exposed to ultraviolet light of a carefully controlled frequency and period of time in the clinic. This is often helpful, but an experienced clinician must supervise this PUVA (psoralens plus ultraviolet radiation) treatment because of the potential hazards. There is no specific diet recommended to ease psoriasis, but many readers have reported benefit from putting curcumin or turmeric on their food. Preliminary research suggests that there may be some scientific basis for their enthusiasm.[1]

FLAXSEED OIL

I've been fascinated with remedies people are using for psoriasis. Several years ago my husband suffered from psoriasis on his arms and legs. The medication the dermatologist gave him was not helpful. I bought some flaxseed oil capsules in the health food store. My husband was skeptical, but by the time he finished the first bottle we saw improvement. He has continued with it and is still free of scabs. I don't know why it works, but others might want to know about it.

A Thanks for sharing your husband's experience. Flaxseed oil, like fish oil, is rich in essential fatty acids, especially alpha-linolenic and linoleic acid. There are reports that flaxseed oil is beneficial for cell membranes and skin health. Although we have not seen any double-blind, placebo-controlled trials, some people report improvement in both psoriasis and eczema through use of flaxseed or fish oil.

LISTERINE

Q My father had a problem with psoriasis of the scalp. A doctor told him to rub a small amount of Listerine (original formula) into his scalp each morning. He does and hasn't had flaking or itching since.

A Many readers have shared their success with Listerine in fighting dandruff. This condition is often caused by a yeast infection. The alcohol and herbal oils in Listerine have antifungal properties that may control the infection. How Listerine might help psoriasis, which is not caused by infection, is a mystery.

TURMERIC

Q A few months ago you wrote about the use of turmeric for boils and possibly arthritis or cancer. This bit of information has changed my life. I've suffered with psoriasis for 25 years and have it over nearly half my body. I've seen many physicians and tried every medication and ultraviolet treatment. The enormous cost has been matched only by my disappointment. When I read that turmeric might have anti-inflammatory properties, I wondered if it might help me. I immediately bought some and sprinkled a rounded teaspoonful on my cereal. I continued the regimen daily, and the results are unbelievable! After ten days, the awful itching and bleeding had ceased. My scalp, which had been heavily flaked and itchy, was returning to normal. The skin problems on my legs and thighs cleared up after eight weeks. Now, five months later, I have no psoriasis, just a few reddened areas where it was bad. I am grateful to you for the information that made a huge difference for me.

A Turmeric, a spice in curry powder, is popular in India. In this country, yellow mustard often contains some turmeric. Curcumin, the active ingredient, has antioxidant and anti-inflammatory properties, which may help explain how it could help an inflammatory condition like psoriasis. Several readers have reported that taking turmeric capsules or putting the spice on food is helpful against psoriasis. Turmeric was used for digestive problems in traditional Chinese medicine.

Q I want to thank you for writing about turmeric for treating psoriasis. I developed this condition two years ago, and it made my skin very itchy and sensitive as well as unsightly. I saw three different dermatologists who all diagnosed psoriasis. Each prescribed creams and ointments, but none worked. After I read your article on turmeric, I tried it. Within one month I was better. After three months, every sore was gone. They have not returned, even though I stopped taking the turmeric nearly a year ago.

A Several readers have reported that taking turmeric capsules or putting it on food helps relieve psoriasis, and one reader has found it helpful for treating irritable bowel syndrome.

Q I want to thank you for writing about turmeric. I had psoriasis on my feet and my hands so bad that I lost all my fingernails. I went to doctor after doctor to heal my psoriasis, but nothing worked. When I saw the article, I thought I'd try it. I put turmeric on my food and in my coffee. Within two weeks, my psoriasis had started to heal. Within three weeks, it had cleared up. I have been free of psoriasis for six months now. My foot is no longer scaling, and the nails on my hands have grown back. I told the doctor about turmeric. I guess he didn't believe me, but I know better.

A Research demonstrates that a component of turmeric, curcumin, shows promise for treating psoriasis and other conditions.[2] We have heard from many others that turmeric can help ease psoriasis. Some people are allergic to the spice, however, and those who take the anticoagulant warfarin should avoid it.

Q My stepdad has had a horrible case of psoriasis for over ten years. The rash was all over his body and caked on his scalp, and it itched constantly. The dermatologist he saw prescribed Dovonex, Capex, and clobetasol. He has used these medications off and on since about 2001, but they provided minimal relief. Mom found a suggestion in your book (*Best Choices from the People's Pharmacy*) that turmeric might help treat psoriasis. We went to the local food co-op and picked it up in pill form. Within a week his scalp was halfway clear, and now, three weeks later, it is just a tiny bit flaky with no itching. This has been life changing for him. Nothing ever cleared up his psoriasis like the turmeric has. He discontinued using Dovonex and wants to drop the Capex and clobetasol next. All of these medications are very expensive, especially in comparison to turmeric.

A People in India have used turmeric in traditional ayurvedic medicine for thousands of years. Basic research suggests a number of reasons why curcumin, the active compound in turmeric, would have a beneficial effect in treating psoriasis.[3] We hope to see a double-blind study of the benefits of curcumin someday. Researchers looking into the potential health benefits of curcumin found that curcumin inhibits an enzyme called PhK, which is associated with overactive cell growth in psoriasis.[4]

Q I know you have written about taking turmeric for psoriasis. My fingernails are falling out from this condition. Is turmeric safe?

A There is growing interest in turmeric and its active ingredient, curcumin, for treating a variety of inflammatory conditions, including psoriasis. Not everyone will benefit from turmeric, and there are cautions. Some people experience skin rashes or liver enzyme elevations. Turmeric may also interact with the anticoagulant warfarin (Coumadin) and increase the risk of bleeding. A new study shows that turmeric increases oxalate in the urine, so it may increase the risk of kidney stones in susceptible people.[5]

THE PEOPLE'S PHARMACY

Favorite Food #23: Curry (Turmeric and Curcumin)

Many spices can go into curry powder, but the one that is absolutely necessary for color and flavor is turmeric. Derived from the root of *Curcuma longa*—a plant related to ginger—turmeric is the golden yellow spice in curry and yellow mustard. It has been widely used in the cuisines of southern Asia for centuries, and lately it seems that the rest of the world has grown to appreciate turmeric as well.

The compounds in turmeric are not easily dissolved in water, but the tradition of making curry with ghee (clarified butter) or coconut milk overcomes this problem by providing fat to carry the important components. Researchers have been studying an ingredient in turmeric called curcumin.[1] This compound has both antioxidant and anti-inflammatory properties that may contribute to its use for treating conditions like arthritis, psoriasis, and gout.

Cancer researchers are investigating the potential of curcumin to affect an extraordinary range of tumors involving the breast, pancreas, colon, bladder, prostate, skin, ovaries, lung, and brain. Some data even suggest that curcumin might be active against the blood cancer multiple myeloma.[2] Research also suggests that the compound curcumin may even be able to ward off Alzheimer's disease, which seems to be rooted in brain inflammation.[3]

We do, however, want to issue a couple of cautions. If you develop a rash or hives, you should stop taking the turmeric immediately. The reaction could worsen. Some people have found that their liver enzymes go up when they take turmeric or curcumin. This is a warning sign that the liver is having trouble.

In addition, we have had several reports that turmeric can interact with the blood-thinner warfarin (Coumadin). If you take Coumadin, you should not take turmeric. It is too difficult to strike a balance between turmeric and warfarin, and the danger of bleeding is too great to risk taking both.

Raynaud's Disease

Raynaud's is a condition in which blood vessels in the hands and feet constrict, leading to pain and numbness. Sometimes fingers or toes even turn white or blue. The nose or earlobes also may be affected. The cause is mysterious, and there is no cure. Doctors prescribe a variety of medications to improve circulation, but no one treatment works for everyone. For some, home remedies may be a worthwhile option.

ASTRAGALUS

Q About a year ago I read in your column that a man with Raynaud's phenomenon had good luck with astragalus root for this problem. I tried it, and within days it had worked like magic. Before using it I had almost decided I would move to Florida for the winter. I didn't take it during the summer, because my fingers become painful only in cold weather. This winter I began taking it, and it is working beautifully again.

A Astragalus root has been used in China for centuries to boost immunity and circulation. Research on animals suggests that astragalus may improve circulation, which is a problem with Raynaud's.

CINNAMON

Q My Raynaud's syndrome has suddenly seemed to disappear. The only change I've made is to take a cinnamon capsule daily starting about six weeks ago. My fingers used to turn white even in the summer. Today it's in the 20s and I haven't had a problem, even getting stuff out of the freezer. Is it the cinnamon?

A In Chinese medicine both cinnamon and the herb astragalus have been used to improve circulation and to relieve symptoms of

Raynaud's disease. Be careful not to overdose on the cinnamon, though, since this spice sometimes contains a compound called coumarin that can damage the liver. To be safe, try looking for a cinnamon supplement that is water extracted to eliminate the coumarin.

FAVORITE RECIPES 

T's Immuno-tea

Contributed by Tieraona Low Dog, M.D.

1 ounce astragalus root, dried and sliced
1 ounce schizandra fruit, dried
1 ounce rose hips, dried
½ ounce eleuthero root, dried and sliced
2 cups water
Honey and lemon to taste

Mix herbs together and store in a jar. Keep in a dark place.

TO PREPARE: Add one tablespoon herb mixture to water, bring to a boil, and simmer, covered, for 15 minutes. Strain. Add honey and lemon to taste.

Drink half a cup two to three times every day during the cold and flu season.

Restless Leg Syndrome

For years, people with restless leg syndrome (RLS) tried to describe their experience as a creepy-crawly feeling under the skin, or a kind of tugging or gnawing in the muscle. Some people called it jumpy legs because the sensations are eased only when they move their legs. The condition is most noticeable when someone is sitting still or lying down. RLS may be accompanied by periodic limb movement in sleep. In this condition, muscles twitch or jerk (every minute, or even several times a minute) all night long and disturb sleep. Prescription drugs can help ease RLS, but side effects such as daytime sleepiness (with potential accidents for drivers) have many sufferers looking for home remedies.

GIN-SOAKED RAISINS

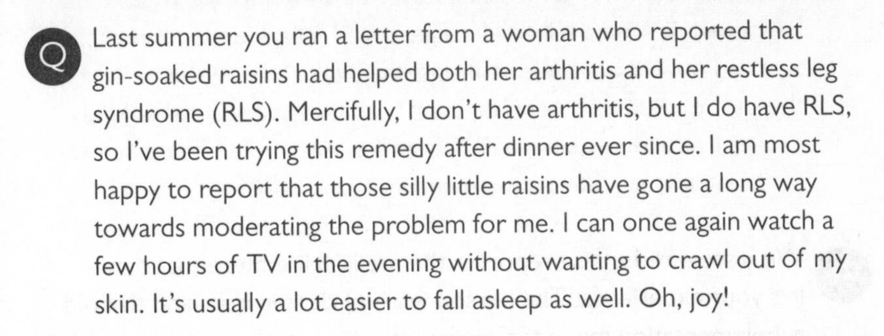

Q Last summer you ran a letter from a woman who reported that gin-soaked raisins had helped both her arthritis and her restless leg syndrome (RLS). Mercifully, I don't have arthritis, but I do have RLS, so I've been trying this remedy after dinner ever since. I am most happy to report that those silly little raisins have gone a long way towards moderating the problem for me. I can once again watch a few hours of TV in the evening without wanting to crawl out of my skin. It's usually a lot easier to fall asleep as well. Oh, joy!

A We have been writing about gin-soaked raisins since 1994. We have no idea why some people find them so helpful for arthritis pain. Last July we received this letter from a reader: "I read about your home remedy of white raisins soaked in gin to help arthritis pain. I tried this and found only a moderate improvement in arthritis pain. But after two weeks of treatment I noticed a marked improvement in RLS. Have others reported this seeming cure? I used to experience RLS two or three times a week, but have not had a recurrence since beginning the gin and white raisin treatment."

IRON DEFICIENCY

Q For a long time I had trouble keeping my legs still, especially while I was trying to sleep at night. I did not seek treatment because I did not want to take medication. Then I read an article that said one of the health conditions causing this was an iron deficiency. My cousin told me she was taking iron twice a day and it helped ease her restless legs. My husband was taking iron pills for a deficiency, so I borrowed a few and tried them. I could tell a difference within two days. I am grateful for such an easy remedy.

A Research has linked iron deficiency to RLS. Correcting this mineral deficiency may help ease the symptoms.[1]

Q I have severe RLS, but it has been controlled with diazepam. Recently the doctor diagnosed me with an underactive thyroid condition and prescribed levothyroxine. It really made my RLS worse. He suggested that I stop the thyroid medicine for three weeks to see what happened. It took a full week to get my RLS back under control. I am worried now about the doctor insisting I take levothyroxine for my thyroid condition. Is there anything I can take instead? I absolutely cannot live with my severe RLS. It affects my whole body, not just my legs, and even affects me mentally.

A We discovered a case report in the medical literature that parallels your experience. The person was deficient in iron, and thyroid supplementation made the creepy-crawly sensations and limb movements worse.[2] Perhaps your doctor can check your iron levels to see if you need a supplement. Untreated hypothyroidism is associated with a number of uncomfortable symptoms, including mental sluggishness, depression, and confusion. It can cause high cholesterol, constipation, fatigue, swollen hands or feet, and weakness, among other problems.

MAGNESIUM

Q I started getting muscle movements in my legs in the mornings while lying in bed. It wasn't restless leg syndrome because I did not feel like I had to move my legs. I just had an unusual feeling of muscles moving under the skin. I heard a discussion on the radio about this that suggested using magnesium. I've started taking 1,000 milligrams before bed, and though it doesn't stop it, it does minimize the discomfort somewhat. Is 1,000 milligrams of magnesium a day too much?

A Research suggests that magnesium may be helpful in easing restless leg syndrome.[3] Your body will let you know if you are taking too much magnesium. In excess, this mineral causes diarrhea. For most people, 300 milligrams a day or so is tolerable, but 1,000 milligrams would not be. People with kidney problems should avoid any supplemental magnesium, as it could put too much strain on the kidneys.

SOAP

Q My husband has had RLS for a long time. I kept reading about the soap remedy in your column, and I finally decided to put the soap in our bed without his knowledge. It worked! It's really unbelievable. One day last month my son was home doing his laundry. He always folds his clothes on our bed, and he came to me saying, "Mom, do I want to know why there's a bar of soap in your bed?"

A We wish we knew why a bar of soap under the bottom sheet often helps ease restless legs. This condition, a persistent, irresistible urge to move the legs, can interfere with sleep. Not only the sufferer but also the bed partner may end up sleep deprived.

Q I have suffered from restless leg syndrome all of my life, and I am over 50! My mother told me I had growing pains when I was a kid, but that became ridiculous when I reached my 20s. When I read about putting soap under the bottom sheet, I thought it was the silliest thing I had ever heard of. Let me assure you that to my great surprise, it works. I have now had five straight nights of sleep in my own bed with no walking around and no shifting to try to get comfortable.

A Putting a bar of plain soap between the bottom sheet and the mattress pad certainly seems bizarre. We have no idea how it may work, but we have heard from many happy readers that it does.

Q I finally fell asleep last night but woke up an hour and a half later. I had started getting cramps and a creepy-crawly feeling in my legs. I know some people who have RLS, so I decided to look it up and see if there were any remedies. I'm sitting at the computer right now with a bar of soap under each leg, and it seems to be working! (I thought this idea was crazy, but I figured I had nothing to lose.) Thank you.

A Placing a bar of soap under the bottom sheet, near where the legs will rest, is one of the wackiest remedies we've come across. But we have heard from so many people that it helps, it is clear some folks do benefit. Soap is inexpensive and doesn't have side effects. That cannot be said about the medications prescribed for RLS.

Q I have severe periodic limb movement disorder (PLMD). I have used Sinemet and Mirapex separately during the last eight years, but I stopped each due to side effects. I am currently unmedicated and miserable. The leg movements at night destroy my sleep, but I am waiting for the doctor to decide what to do next. Someone mentioned Ivory soap, so I tried it in a wrap around my ankles and under my feet in double socks. I have also placed four bars across my lower bed under the bottom sheet. No matter where I move my feet, a bar of soap is near. If a movement wakes me, I move my foot to touch the soap, and fall right back to sleep. This hasn't been a cure, but it certainly has made the limb movements more tolerable. I would love to understand why it helps, but just having a degree of relief is good enough for me.

A The medications you have taken, along with one called Requip, are sometimes prescribed for PLMD or for a milder form called restless leg syndrome. As you discovered, the side effects can be hard to tolerate. We worry particularly about people falling asleep during normal daily activities such as driving. Yikes! In that context, soap begins to sound pretty good. There are no studies, and no one knows why it might help. So many readers have told us that they have gotten relief from RLS or leg cramps by using soap under the bottom sheet, we think it is worth a try. Even PLMD might be less troublesome. And soap won't put you to sleep behind the wheel!

Sex/Libido

Human sexuality is tricky business. Sex involves the brain, blood vessels, hormones, and nerves and muscles. Romance and relaxation, as well as other factors, affect libido and sexual fulfillment. We have drugs like Viagra and Cialis, but they do nothing for libido and don't always lead to successful sex. A surprising number of people want suggestions to dampen desire. Sometimes a natural approach can help.

MUIRA PUAMA

Q My husband and I are in our 70s and have been married forever. I have had a problem with vaginal dryness. My husband read about muira puama and bought some online. I take it as directed twice a day, and we make love at least five times a week.

A According to one report, this Amazonian herb has a reputation for treating sexual difficulties, including low libido and erectile dysfunction.[1] However, there is very little scientific data about its safety and effectiveness. Side effects may include headache, digestive upset, and nervousness. We could find no research on the use of muira puama for vaginal dryness. Your experience is certainly intriguing.

TESTOSTERONE

Q What do you know about compounded testosterone cream? I am a 64-year-old woman with low libido. A friend of a friend uses this cream before sex by applying it to the inner thighs. She has great results with desire and orgasms, but my gynecologist says it is not approved by the FDA and won't prescribe it.

A Low testosterone levels in men and women are associated with diminished sexual interest, arousal, and enjoyment. Some studies suggest that testosterone therapy may boost libido, even in women.[2] The dose is important: Too much of this male hormone can cause facial hair growth, acne, deepening of the voice, and clitoral enlargement. Your doctor is correct that the FDA has not approved testosterone for improving women's sex drive. But a physician specializing in sexual medicine may be able to assist you.

VITEX

Q I have a sex drive more intense than my partner's. He is in his late 30s, and I am in my mid-20s. I would prefer to have sex every couple of days, while he is fine with having sex once every week and a half. I find it difficult to deal with this situation. I worry that I might make sex seem like a chore to my partner. I often wish that I could take medication to diminish my sex drive so that I could be happier.

A We consulted two leading sex experts about the concerns you raise. Dr. Ruth Westheimer suggested your partner could help you achieve satisfaction even if he isn't in the mood for intercourse. Irwin Goldstein, M.D., editor of the *Journal of Sexual Medicine*, pointed out that you are not unusual. In focus groups, 25 percent of women report having a higher sex drive than their male partners. Your partner may want to have a medical workup and have hormonal levels checked. However, the herb vitex (chaste tree berry) may reduce sex drive.

YOHIMBINE

Q I am a 66-year-old male in very good health. My wife and I have relations twice a week. My problem is difficulty in reaching orgasm. I satisfy her just fine but not myself. This is frustrating. I am in good shape, work out regularly, take supplements, and am on natural testosterone replacement. I do take Diovan daily for hypertension. Could this be the problem? Do you have any suggestions?

A It doesn't seem likely that the Diovan is to blame. The only study that addressed this question found that Diovan improved sexual function.[3] An old-fashioned herbal remedy may be helpful. Yohimbine is derived from the bark of an African tree. In a recent study, this herbal extract restored orgasm in about half the men treated.[4] A doctor familiar with its use should supervise treatment, especially since yohimbine can raise blood pressure. Other potential side effects include palpitations, anxiety, dizziness, and digestive distress.

Shingles

Anyone who had chicken pox as a child may develop shingles at some point as an adult. The varicella-zoster virus that causes chicken pox goes into hiding and only reemerges when something, such as stress, weakens the immune system. Shingles can cause extreme tenderness on just one side of the torso or face. The rash or blisters may appear a few days later. Even after the rash clears up, the person may be left with lingering chronic pain. Prescription drugs can treat shingles successfully, but they work best if they are taken in the first few days of the attack. A person who misses that window of opportunity may be interested in home remedies.

APITHERAPY

Q My dad has suffered with pain from shingles for almost two years. Yesterday, during a nap on the deck, he got stung on the toe by a bee. Now he's feeling no shingles pain! Have you heard of such a thing? He's perplexed but elated, for however long the freedom lasts.

A Years ago we received this letter: "While snoozing on the porch I was stung on the finger by a tiny bee. The result: intense pain, and after that a great reduction of arthritis in my arm." Some prominent doctors used bee venom therapy to treat arthritis in the 1920s and 1930s. Hospital pharmacies even stocked injectable bee venom. Apitherapy fell out of favor due to lack of scientific proof. But people are now trying bee stings again for arthritis and postherpetic neuralgia (pain lingering after shingles), as well as for other conditions. Some people keep honeybees to sting their sore joints. Others get their doctors to inject venom used for bee allergy desensitization. No one who is allergic to bee stings should try this treatment, for one sting could trigger a fatal reaction.

GREEN TEA

Q I suddenly developed a severe case of shingles around my waist on the right side. I didn't want to get involved with strong pain relievers, so I tried applying green tea four times on the small blisters that formed the first day. The next day, after applying green tea again four times, the scabs fell off, and I have had no lasting pain. Perhaps this will help other people suffering from shingles.

A Shingles is an extremely painful skin eruption caused by the same virus that causes chicken pox (the varicella-zoster virus). Prompt treatment with prescription antiviral medicines such as acyclovir (Zovirax), famciclovir (Famvir), or valacyclovir (Valtrex) can speed healing. One should never rely on home remedies for such a potentially serious condition as shingles, but your experiment is fascinating. Applying tepid green tea in addition to taking appropriate medical treatment should not pose a risk. Tests on animals have shown that topical use of green tea has anti-inflammatory benefits.

LISTERINE

Q What is the miracle of Listerine? Twenty years ago I got shingles. I had a painful, blistery rash. My doctor told me to rub Listerine on it. The itching stopped, the rash disappeared, and the pain went away. I also had gums that bled every time I brushed my teeth. The dentist told me to dip my toothbrush in Listerine and apply to my gums after brushing. The problem vanished, and I haven't had trouble since.

A Listerine contains a number of herbal extracts (thymol, eucalyptol, menthol, and methyl salicylate) in an alcoholic solution. There are no studies to suggest that such ingredients could help relieve shingles or speed healing from gum disease. Nevertheless, one or more of the oils in Listerine might be beneficial for such conditions.

Sinusitis

The sinuses (there are several pairs) are air pockets in the bones of the skull. (Yes, we all have holes in our heads.) They connect to the respiratory tract, and when the nasal passages get inflamed, the sinuses may be involved (sinusitis means "inflamed sinuses"). This inflammation may also be triggered by seasonal allergies or infection. A sinus infection might need antibiotic treatment, but home remedies may reduce the inflammation.

SALINE SOLUTION

Q I just read your column in which a person with sinus problems advocates using a saline solution twice a day. I have been doing this for years when my sinuses start to act up. The only difference is I make my own saline with water and table salt.

A One recipe for saline solution to wash nasal passages calls for one-quarter teaspoon salt to eight ounces of tepid water.

NETI POT

Q When I was a child, my mother told me to gargle with salt water for a sore throat. I just heard Dr. Oz on Oprah recommend salt water in a neti pot for improving sinus conditions. Are neti pots safe to use?

A A neti pot looks a little like Aladdin's lamp. You put lukewarm salt water in it, hold your head upside down, and pour the solution into one nostril until it runs out the other. This ancient Indian practice helps wash out the nose and sinuses. It should be safe. But you can achieve the same effect with a saline rinse found in most pharmacies.

Skin Fungus

Skin fungus is extremely common, especially in the hot, humid summer months. Warm, dark places are particularly susceptible to fungus. But it's possible to develop athlete's foot, ringworm, and other itchy, unpleasant conditions at any time of year. They're contagious and spread on the floors of wet gym showers, at nail salons, or from contact with infected skin. An ounce of prevention goes a long way, so consider taking a pair of flip-flops along on your next visit to the gym shower. But no matter how ubiquitous the fungus, there are remedies to try if you wind up with the telltale itch.

ANTIPERSPIRANT

Q My husband suffered from jock itch for a long time. He would use medicine to clear it up, but it always returned. Then he tried applying antiperspirant to the affected areas daily. This solved the problem.

A Jock itch is caused by a fungus that thrives in moist areas. The antiperspirant probably keeps his skin dry and discourages fungal growth.

LISTERINE

Q I have been plagued with jock itch for weeks and had tried a couple of OTC creams with little success. I was getting ready to see my dermatologist when I read about using Listerine for jock itch, athlete's foot, and other fungal infections. Listerine has now cleared the problem up in a couple of days. It stings for a minute or two when first applied, but it isn't that bad and it really worked.

The herbal oils in the original formula Listerine include eucalyptol, menthol, and thymol. There is some evidence that these herbal extracts in combination may have antifungal properties.

MILK OF MAGNESIA

Q

A friend who is an internist recommended a mixture of milk of magnesia (MoM) and Lotrimin AF to combat seborrheic dermatitis on my face and the backs of my ears. She suggested mixing roughly half a 12-ounce bottle of MoM with a whole tube of the Lotrimin AF cream. The first application certainly had a positive effect on my skin. I did not follow through as I should have, so I don't know how well it works in the long term. Have you ever heard of this remedy?

A

We could find no research on this intriguing remedy for seborrheic dermatitis. This skin condition is characterized by itching, flaking, scales, and redness. It frequently occurs on the scalp as superdandruff or even on the eyebrows, on the forehead, around the nose, or on the chin. This condition appears to be an inflammatory response to fungi that belong to the genus *Malassezia* and are known as yeast. Dermatologists frequently treat this problem with antifungal creams (such as clotrimazole, the active ingredient in Lotrimin AF). Topical steroid creams such as hydrocortisone are also used. Dandruff shampoos containing ketoconazole, selenium sulfide, or zinc pyrithione can be helpful. Readers claim that applying milk of magnesia to the armpits is a gentle and effective way to reduce sweating and odor. Perhaps the drying effect and alkalinity of MoM together with the antifungal activity of Lotrimin AF discourage yeast belonging to the genus *Malassezia*.

Q

I have been using milk of magnesia on my face for the past two months since reading about it in your column. My face flakes are gone! I pour it in my hand and massage it on my face—forehead, eyebrows, and around the eyes, nose, cheeks, and chin—while showering. Then I rinse it off at the end of the shower. It's a great, cost-effective alternative to expensive Nizoral, and it works better.

A reader told us that a doctor suggested a topical mixture of milk of magnesia and Lotrimin AF for seborrheic dermatitis.

Skin Tags

These fleshy little benign skin growths can show up in armpits or groins or on the neck, eyelids, or face. They may range from the size of a grain of rice to the size of an almond. Doctors aren't sure what causes skin tags, though people who are overweight or have diabetes seem more prone to developing them. Women may also discover skin tags during pregnancy. A dermatologist can remove skin tags without much fuss, but some may want to consider home remedies.

LIQUID BANDAGE

Q In one of your articles you stated that a reader used New-Skin Liquid Bandage to remove skin tags. I would appreciate it if you would address this again and describe how New-Skin was used. I recently saw a dermatologist, and he wanted $300 to remove about 12 small tags.

A A few years ago we heard from a reader who had managed to get rid of skin tags by covering them tightly with a Clear Spots Band-Aid. Several months ago another reader reported that he had tried the special Band-Aids but "could never get a bandage to stay on long enough." He was about to give up when he ran across some liquid bandage in his medicine cabinet. He told us, "I had a large flap growing on my shoulder and put the New-Skin Liquid Bandage on it. Within a week the flap fell off. I put it on some smaller skin tags, and they shriveled and fell off too." This reader provided no clear instructions. But we have heard from many people who have applied liquid bandage one or two times daily with good results. One wrote, "New-Skin for skin tags worked for me too! I did reapply the product several times, and they did shrink and were pulled off when removing the 'bandage' after about ten days. This saved me quite a bit of money."

Q Some time ago I read in your column about someone who had success removing skin tags with a liquid bandage. I would appreciate hearing about this remedy. I have several of these growths around my neck where the chain of my necklace rests. I have been considering having them removed by a dermatologist, but would like to try this remedy first.

A We have heard from many readers who tell us New-Skin Liquid Bandage works: "I used it twice a day for three days, and the skin tags came off!" Others tell us it may take a few weeks. Another approach some readers have tried involves a wart remedy: "I found Compound W works just as well for getting rid of skin tags."

THREAD

Q I have some skin tags in my armpits. Is there anything to put on them so they will go away? I don't want to pay a doctor to cut them off.

A One approach requires some fortitude: "I am a nurse, and for years I have tied a piece of thread around the tag at the base, pulled it tight, made a tight knot, and cut off the long ends. (It stings at first, and then it's all right.) After three or four days the tag turns black and falls off, similar to bobbing a lamb's tail. It strangles the blood supply and works every time. It helps to have someone do it for you." Another reader reported a similar experience: "I had a rather large one on the side of my neck. My dermatologist said to just ignore it. Soon after, I mentioned to a friend (an orthopedic doctor) how much I hated it, and he promptly tied and knotted a piece of thread around it and cut the ends off real close. You couldn't even see it. He said it would fall off very quickly. It was gone in three days."

Stinky Feet

Bacteria, fungi, and goodness knows what else can thrive between your toes and inside your shoes. For reasons that are not entirely clear, some people have the smelliest feet imaginable. We have collected all sorts of fascinating remedies for this condition.

ANTIPERSPIRANT

Q My 12-year-old daughter is a ballet dancer and has started pointe. Her feet smell so bad we gag if she takes her shoes off. Do you have any remedies for foot odor?

A Foot odor seems to be a common problem among young ballerinas. The mother of a 20-year-old dancer offered this advice: "First, get some shoe dogs. These are cedar-filled bags that absorb the moisture in the shoe and help with the odor. Second, ballet students also wear classic soft ballet slippers. Canvas slippers are better than leather ones, since the canvas kind of slipper can be washed every other week if the need arises. With daily ballet classes, shoes don't dry out, so purchasing a few pairs will help. They should be stored in mesh bags, not plastic, and outside the dance bag, not in it. Third, you might try a dry, rub-on antiperspirant on the feet once a day. This also helped my son with his sweaty, smelly soccer feet. Fourth, if she is new to pointe, she may be wearing pads in her ballet shoes to protect her toes. She should use natural lamb's wool pads that allow the skin to breathe."

BAKING SODA

Q My son-in-law and grandson have the worst smelly feet. They can clear a room in a manner of minutes. Actually, all the men in their family suffer from this. Help!

A Soaking the feet in a baking soda or in an Epsom salt solution may be helpful. Some people report that chlorophyll tablets or zinc can banish the odor. Your relatives should be cautious about zinc, though. Too much can interfere with the body's balance of some other minerals.

LISTERINE

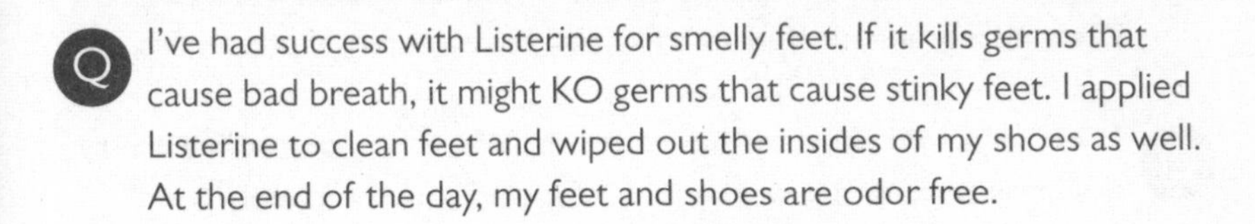

Q I've had success with Listerine for smelly feet. If it kills germs that cause bad breath, it might KO germs that cause stinky feet. I applied Listerine to clean feet and wiped out the insides of my shoes as well. At the end of the day, my feet and shoes are odor free.

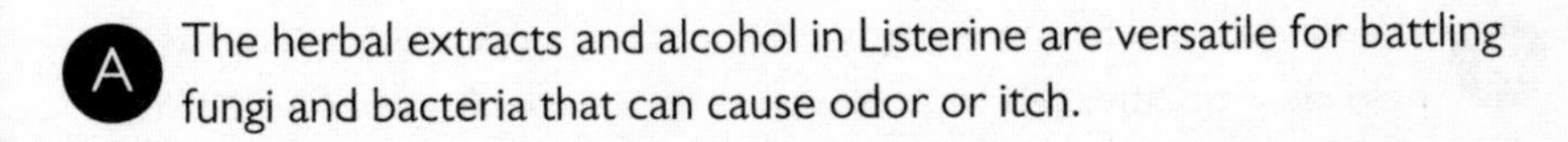

A The herbal extracts and alcohol in Listerine are versatile for battling fungi and bacteria that can cause odor or itch.

TEA

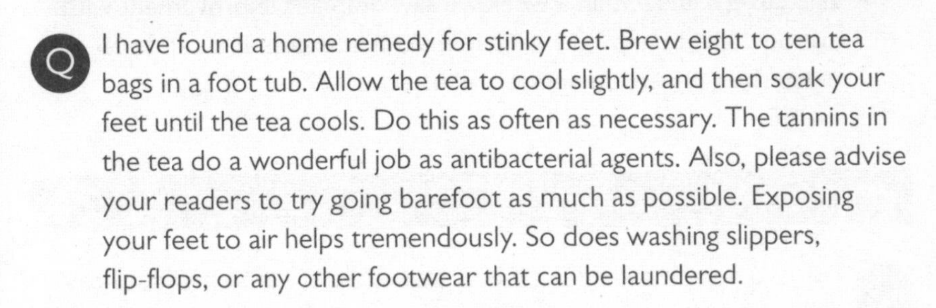

Q I have found a home remedy for stinky feet. Brew eight to ten tea bags in a foot tub. Allow the tea to cool slightly, and then soak your feet until the tea cools. Do this as often as necessary. The tannins in the tea do a wonderful job as antibacterial agents. Also, please advise your readers to try going barefoot as much as possible. Exposing your feet to air helps tremendously. So does washing slippers, flip-flops, or any other footwear that can be laundered.

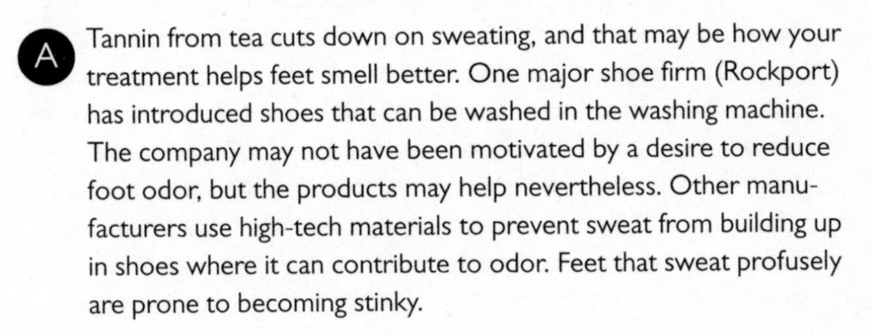

A Tannin from tea cuts down on sweating, and that may be how your treatment helps feet smell better. One major shoe firm (Rockport) has introduced shoes that can be washed in the washing machine. The company may not have been motivated by a desire to reduce foot odor, but the products may help nevertheless. Other manufacturers use high-tech materials to prevent sweat from building up in shoes where it can contribute to odor. Feet that sweat profusely are prone to becoming stinky.

URINE

Q I read that if you had really bad foot odor that all you had to do was urinate on your feet to make it go away. I tried this, and not only did the odor go away, but the bad painful peeling of the skin on the bottom of my feet went away also. Why would that happen?

A Soldiers have used this military secret for decades to treat foot fungus and odor. Perhaps the acidity of the urine does the trick. Urea, an ingredient of urine, may also have antifungal properties.

Q My 17-year-old daughter has very smelly feet. I convinced her to try the urine trick, and it worked! Her foot odor is completely gone!

A We have heard from veterans that urinating on smelly feet can help clear up athlete's foot as well as foot odor. Another possibility involves zinc supplements. Here is one reader's story: "About eight years ago, my then-12-year-old son had terrible foot odor. It finally stopped when I read in your column to try zinc. We bought generic zinc tablets, and within a week we saw dramatic improvement with just one tablet daily." Do keep in mind that excess zinc can have deleterious effects on nerves and muscles.

VINEGAR

Q For two decades, I had the worst case of smelly athlete's foot I have ever seen. That's really saying something, because I'm a doctor. I've seen (and smelled) many. The fungus also caused deep fissures in the soles of my feet and between my toes. I tried everything: griseofulvin (Grisactin), gentian violet, Clorox, Absorbine Jr., tolnaftate (Tinactin), Desenex, and white cotton socks. In the early 1990s, my mother suggested white vinegar foot soaks. I bought a gallon of vinegar and a small basin and began soaking my feet twice daily. I dried them without rinsing the vinegar off, then put on socks dusted with Desenex powder. At first the fissures stung, but the itching and smell began to fade almost immediately. I continue to use this regimen twice or thrice weekly. Soaks of five to ten minutes are long enough.

Your story is very convincing! We've heard from other readers who've had success treating athlete's foot with vinegar.

Sunburn and Sun Rash

Many dermatologists see the sun's rays as the enemy. Ultraviolet radiation causes premature aging, wrinkling, and skin cancer. But there is growing recognition that the vitamin D your skin makes when exposed to sunshine is essential for good health. A little sun exposure (10 to 15 minutes) three or four times a week may be a good thing. Too much sun, however, can wreak havoc. We always encourage common sense. Here are some home remedies to help if prevention fails.

LISTERINE

Q For over 25 years I have been using Listerine full strength in a spray bottle for sunburn. An old fisherman told me about it when I got a sunburn from sitting in a boat on a lake. I got burned so badly I looked like a lobster. The people I tell say this remedy works for them too. It stops the pain instantly. I keep it in my travel bag, in a zipper-top plastic bag to avoid spilling. I don't like Listerine as a mouthwash, but it's great for sunburn. You don't have to touch the skin and hurt yourself even more while you're applying it.

A This is a fascinating use for Listerine we've not encountered before. Perhaps the menthol or eucalyptol in the original flavor Listerine have cooling properties. Thanks for sharing your remedy. We hope, however, that you don't have many opportunities to use it. Dermatologists tell us that multiple sunburns increase risk of melanoma and other skin cancers. Why not put sunscreen in that plastic bag along with the Listerine, so you've got prevention as handy as the cure?

Q I often get a very itchy sun rash when I go on holidays. Listerine calms it down so I can sleep and soon clears it. Listerine also soothed an allergic reaction after a massage. I don't travel without it.

A We always encourage people to avoid excessive sun exposure. Spring and early summer are especially dangerous because people are pale after the winter. An itchy sun rash is different from sunburn or an allergy. You may be suffering from polymorphous light eruption (PMLE), a hypersensitivity to ultraviolet light especially common at this time of year. A dermatologist should check you out to make sure you don't have a more serious autoimmune condition such as lupus. Thanks for sharing your Listerine strategy for this condition.

VITAMIN C

Q Instead of sunscreen, I take megadoses of vitamin C, which has protected me from sunburn for over 20 years. (Of course, I don't tan or freckle either.) I take three grams of C each day, and once every year or two I might get a little pink on the most sensitive areas (tip of my nose, neck, and shoulders early in the summer); otherwise, the C protects me from the radiation of the sun.

A There is some data to suggest that vitamin C may have modest effects against ultraviolet radiation. One study demonstrated that a combination of antioxidants like vitamins C and E could reduce DNA damage from sun exposure.[1] Do not assume, however, that oral vitamins can protect you from the sun's rays. Stay out of the midday sun and also use sun blockers that contain titanium dioxide or zinc oxide.

Vaginal Dryness

This is a common condition after menopause, and it can make marital relations quite uncomfortable. Physicians sometimes prescribe an estrogen cream or ring, but many women would like to minimize their exposure to estrogen.

Over-the-counter personal lubricants can possibly help, but there are also some inexpensive home remedies that seem to work.

COCONUT OIL

Q You had a reader who wanted to know what to do for vaginal dryness. I'd like to respond. From my experience, coconut oil is best. It is inexpensive and widely available at health food stores. It comes in a glass jar. Even though it is a bit solid in the jar, when it is allowed to warm to room temperature it easily dissolves into the skin. It is harmless to the tender tissues of the vagina and has antiviral and antibacterial properties that are very helpful.

A Thanks for the recommendation. We have heard from several women who use olive oil for vaginal lubrication. As coconut oil is edible, it too should be safe. Keep in mind that oil of any sort is incompatible with latex and should not be used with diaphragms or condoms.

CORN HUSKERS LOTION

Q I am a 55-year-old female suffering from hot flashes, vaginal dryness, and loss of libido. Sexual intercourse is uncomfortable. I used to look forward to making love with my husband, but it is hard to enjoy anything that hurts. We have tried drugstore lubricants, but we are not satisfied with any of them. Any suggestions?

A You will have to experiment with personal lubricants. What works for some couples is unacceptable for others. One woman wrote to say that she suffered from dryness starting in her 40s: "I tried numerous treatments, including expensive lubricants. Then I read about Corn Huskers Lotion. The results with this inexpensive hand moisturizer have been incredible." Others tell us that olive oil or the gel from an aloe plant can be used as vaginal lubricants. Sylk is another natural option. It contains kiwifruit vine extract and is available on the Web at *www.sylkusa.com* or by calling 866-406-SYLK.

OLIVE OIL

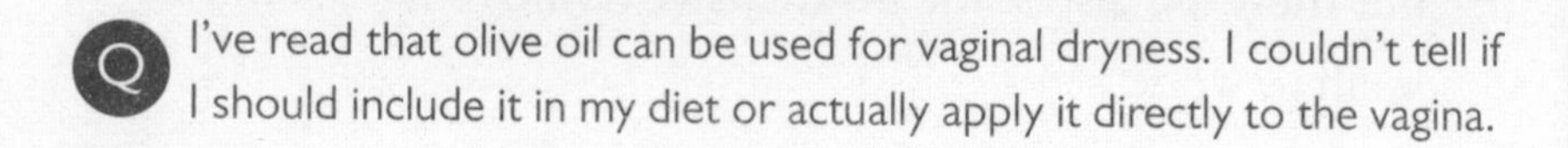

Q I've read that olive oil can be used for vaginal dryness. I couldn't tell if I should include it in my diet or actually apply it directly to the vagina.

A This remedy calls for topical application of olive oil to the vagina. It seems to lubricate those dry tissues. Here is what one reader reported: "I have been using olive oil for this purpose for a couple of years. When my doctor did a pelvic exam, he thought I was taking hormones, although I am not. I think olive oil has natural vitamin E to keep women youthful."

Warts

Warts are noncancerous skin growths that commonly occur on the hands and feet. They are the body's response to a viral infection with human papilloma virus. The virus causes keratin, a hard protein occurring in the top layer of the skin, to grow too fast and make rough, blisterlike growths. Plantar warts on the soles of the feet can be much more difficult to treat. Over-the-counter preparations as well as dermatologists' interventions like cryotherapy (using extreme cold to freeze and destroy tissue) or surgery all work, but many people find that home remedies are successful too.

BACON FAT

Q I read about someone who wanted to know how to get rid of plantar warts. My grandson had one on his foot for quite some time and was going to a doctor for treatment. His grandpa had him put a piece of raw bacon fat on the wart overnight, and in a couple of nights it was gone. To try this, use a fresh piece of bacon each night and cover it with a bandage.

A Plantar warts (small lesions that grow on the soles or toes of the feet) are notoriously hard to treat. Readers have suggested duct tape, castor oil, and hot water soaks. Bacon fat is a new remedy for our collection.

BANANA PEEL

Q I have heard that a banana peel can be helpful against warts, but I am not clear on how you would use it. My doctor burns off several warts yearly, but inevitably they come back. I would be very grateful if you would share specific information on how to use a banana peel to treat warts.

A Cut a piece of banana peel slightly larger than the wart. Use tape to hold the white fleshy side next to the wart. Leave the banana peel on overnight. Use a new slice of banana peel daily for a couple of weeks. We make no promises, but other readers tell us that the banana peel remedy can work. One of our readers complained that burning warts off was both expensive ($600) and painful. Although it took an entire month, the banana peel remedy was painless and affordable. Besides that, the reader noted, she enjoyed eating the bananas as a bonus.

Q We gave our six-year-old daughter a heartburn medicine, cimetidine, for her warts. It's amazing! After months of visits to the dermatologist, the warts on the back of her hand are gone. She had up to 40 big and tiny warts, and they were starting to spread to her wrist and other hand. Finally, we gave her cimetidine daily for eight weeks and they just disappeared.

A The cimetidine (Tagamet) "cure" for warts was first written up in the early 1990s. This was an unusual use; Tagamet was a popular prescription drug for ulcers at that time. Since then a number of studies have tested such acid-suppressing drugs against warts. Although some research subjects had a good response like your daughter, most of the well-controlled trials showed no benefit over a placebo.[1]

DUCT TAPE

Q I had a wart near my ankle and decided to experiment with the duct tape treatment. I cut a small piece each morning, and after my shower I stuck it over the wart. I repeated this every day for three weeks. I noticed the top layer of the wart seemed to come off each time I removed the tape. When the wart was quite smooth, I filed it gently with an emery board and it bled a little. Then I forgot about it. Three weeks later, I looked for the wart and it was gone!

A Controlled studies of home remedies are rare, so it is often hard to assess their effectiveness. In the case of duct tape, though, there have been a couple of trials. In one, the investigators found that duct tape worked better than freezing warts off in children.[2] More recently, scientists reported that duct tape is only slightly better than a placebo when used on schoolchildren[3] and no better than a placebo for adults.[4] Although research suggests that you were fortunate, we have heard from other readers who also rid themselves of warts with duct tape and patience.

GARLIC

Q I've had good luck getting rid of warts with minced garlic. Apply it directly to the wart and cover it with a bandage. Change the garlic and the bandage at least once a day. The wart should be peeled or filed down regularly and will slowly disintegrate. This usually takes about three weeks. Be very careful to cover only the wart, since garlic is too strong for healthy skin.

A Thanks for the suggestion. Home remedies can be useful for warts.

LISTERINE

Q My daughter had several plantar warts on the sole of her foot. We treated them with duct tape for several months with limited success. Listerine has been used for eliminating fungus, so I thought it might

THE PEOPLE'S PHARMACY

Favorite Food #24: Garlic

For those of us who are garlic lovers, there is nothing so wonderful as the smell of garlic being sautéed in olive oil.

Garlic is one of the oldest medicines. It is mentioned in the Bible and the Talmud. Garlic was hung from the doors of homes and shops to ward off potential evils. Ancient Greek and Egyptian healers used garlic medicinally. Hippocrates, the father of modern medicine, prescribed garlic for a number of conditions, from chest pain and leprosy to toothaches and bronchitis. Ancient medical texts from China and India also refer to the healing powers of garlic.

Check the National Library of Medicine (PubMed) database, and you will find thousands of research articles delving into the biochemical and medicinal effects of garlic. Debate rages over the benefits of garlic pills versus fresh garlic.

We won't attempt to lay out the rationales or to take sides, but amid the arguments, we did discover a fascinating study that provides interesting data.[1] Investigators at the University of Connecticut gave rats either freshly crushed garlic or processed garlic. Although both preparations provided heart benefits, the fresh garlic provided "superior protection." Personally, we prefer the real thing to garlic powder, garlic pills, garlic juice, or any other processed form of garlic.

Although data are contradictory about the benefits of garlic for preventing or treating the common cold, recent evidence suggests garlic may indeed be an immune system booster.[2] Other potential health benefits of fresh garlic may include lowering blood pressure and keeping blood platelets from sticking together to cause clots.

Blood vessels may become more flexible after you consume food that contains fresh garlic. Perhaps the most exciting research regarding garlic comes from China, where studies have shown that people who regularly consume garlic have a lower risk of developing stomach or colon cancer.[3]

also kill wart viruses. I put undiluted Listerine in a zipper-top plastic bag and had her soak her feet for ten minutes. She repeated the treatment only a couple of times. The warts disappeared in about three weeks and haven't returned. It might be a coincidence, but she is happy to be wart free!

A Warts are susceptible to a surprising range of home remedies. Readers have reported success with duct tape, a few drops of fresh lemon juice, iodine, and castor oil. Thanks for sharing the Listerine idea. We're glad it worked.

POTATO

Q I had a wart on my finger for 15 years. A few doctors said they could cut if off, but I don't like the idea of surgery. My mother told me of an old gypsy remedy, but I would never try it. It was too hokey. One day I argued with her and, in an effort to prove her wrong, I decided to go ahead and use the remedy. I just knew it wouldn't work. Well, within two weeks the wart was gone. Here it is: Cut a slice from a potato, rub the white part on the wart, and then go bury the potato in the yard. Supposedly if you dig up the potato later, it will have grown the wart on it. This sounds way too silly, but it did work. Mom was right, and she won the argument.

A We don't know if this is a gypsy remedy, but it certainly is old. We have heard from a number of other people who have treated their warts successfully by rubbing them with raw potato or potato peel and burying the piece of potato.

TURMERIC

Q My daughter had three stubborn warts on the bottom of her foot that hurt her terribly. We tried salicylic acid tape from the pharmacy. It always burned off all the good skin and left the wart intact. Duct tape didn't work, and neither did freezing the warts off at the doctor's office. They kept coming back. Then I bought turmeric root from the health food store and scraped off some. I made a paste of freshly scraped turmeric root, taped a little lump the size of a pea on each wart, and changed it each night. (I had her wear a sock to bed. Turmeric is neon yellow and stains bedding.) The warts were completely gone in three applications. Bright pink, fresh, smooth skin grew in the black-specked holes left behind after the bandage was removed. The warts never came back. I have no idea why it works, but someone at work said it worked for him.

A No one knows much about why any wart remedy works. We do know warts are caused by a virus. Studies demonstrate that curcumin, a compound found in turmeric, exhibits antiviral activity, so perhaps that explains your success.[5]

Q My son developed a wart on the bottom of his foot. He didn't take care of it as I told him to, and it grew and multiplied until walking was painful. Nothing he tried helped. His doctor wouldn't touch it and recommended a visit to a specialist. I told him to get some fresh turmeric, but all he could find was the powder. I suggested he make a paste with a little oil. Mixing it with a few drops of olive oil and covering it with a bandage took away the warts and the pain in a couple of days! It has been several weeks, and the warts have not returned.

A You are the second person who has reported success with turmeric for plantar warts. Turmeric's antiviral properties may be the reason.

VINEGAR

Q My daughter is 23 and has had warts on her thighs all the way to her knees for five years. She couldn't wear shorts or bathing suits because of this condition. She had over 50 warts at all times. I lost track of how much money we spent at dermatologists as we tried every treatment available: liquid nitrogen, Tagamet, chemotherapy, and even yeast. A new dermatologist told her she was out of options. The only thing left was to boost her immune system. I heard your radio program and checked your book for possible home remedies. I told my daughter about the vinegar treatment referenced in your book. She wet a paper towel with vinegar and wiped her warts with it twice a day. Her warts are completely gone. I cannot begin to express our appreciation for the information you provide!

It is wonderful to hear that vinegar was so effective.

Weight Loss

Everyone would like a simple way to drop 10 or 20 pounds without working too hard at it. That explains the perennial popularity of diet books as well as over-the-counter preparations (often containing caffeine and/or herbs) purported to control appetite. There have been only a few studies pitting the popular (low-carb) Atkins diet against the (low-fat) Ornish diet. These studies show that sticking with either diet can help dieters lose weight. What's hard is sticking with it! Keeping a food diary may be one of the most helpful "remedies" in a weight loss plan, but here are a few more.

LOW GLYCEMIC INDEX DIET

Q I used to get urinary tract infections or yeast infections every other month. Then I changed my diet and cut out sugar, white flour, and starches like potatoes and rice. Since then I have had only one urinary tract infection. I've lost 20 pounds, and my eczema is 99 percent better. I only have a flare-up when I have cake or milk chocolate. I am still surprised that diet can have such an effect on the system. Other people with eczema or seborrheic dermatitis might benefit as I did.

A There is not much research linking a high-carbohydrate diet to urinary tract infections or eczema. But reducing sugar, starch, and refined carbohydrates seems like a simple enough experiment. If it works for some people with such hard-to-treat conditions, it might be worth the trouble. Thanks for sharing your interesting story.

Q I read that high-fructose corn syrup raises triglycerides, part of the cholesterol count. I eliminated corn syrup from my diet, and my triglycerides dropped significantly. But why did I lose 15 pounds and 3 inches from my waist with no dieting?

A Research has shown that a high-fructose diet can boost triglyceride levels in men. High-fructose corn syrup is found in soft drinks, breakfast cereals, snacks, and other processed foods. Experts have proposed that high-fructose corn syrup in soft drinks and fruit juices is contributing to the obesity epidemic.[1] By eliminating it from your diet, you apparently chose alternatives that helped you lose weight.

Wound Care

Treating and dressing serious wounds is not a do-it-yourself home project. Whether they're the result of a sudden traumatic injury or occurred gradually over time (in the form of a bedsore, for example), you should always be sure to have a doctor oversee the healing process. That said, people have used honey as a wound dressing for hundreds of years, and we've heard from several health care specialists who have used sugar or honey to successfully promote healing in stubborn wounds. There are also some data in the medical literature to support the usefulness of this method. Researchers in Egypt applied honey to 30 diabetic foot ulcers, and after three months of application, the honey completely healed the wounds in more than 43 percent of cases. In another 43.3 percent of cases, honey reduced the size of the wounds and helped with granulation.[1] If you're having trouble with a wound that refuses to heal, talk to your doctor about the possibility of trying a honey or sugar treatment.

HONEY

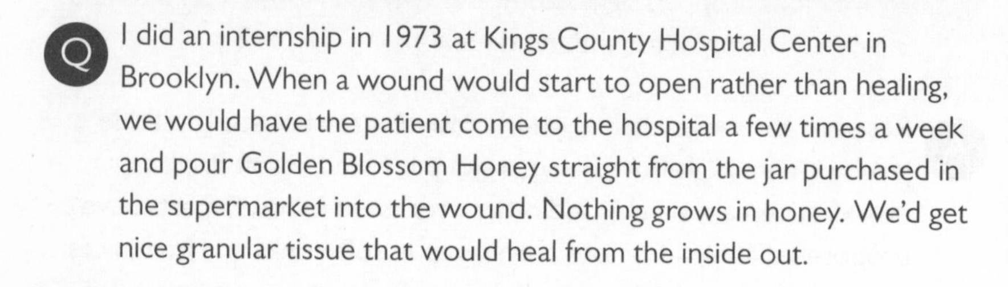

Q I did an internship in 1973 at Kings County Hospital Center in Brooklyn. When a wound would start to open rather than healing, we would have the patient come to the hospital a few times a week and pour Golden Blossom Honey straight from the jar purchased in the supermarket into the wound. Nothing grows in honey. We'd get nice granular tissue that would heal from the inside out.

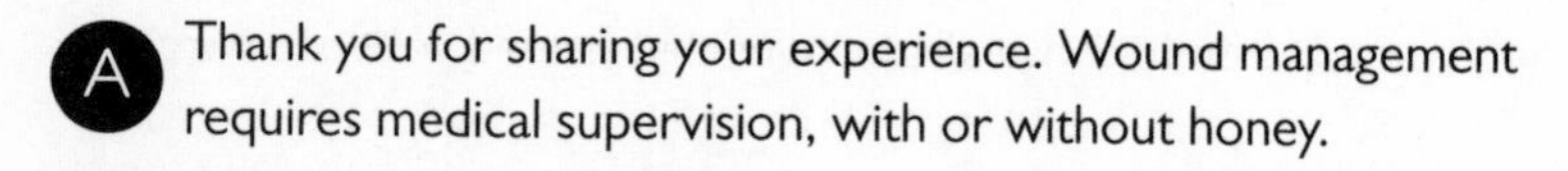

A Thank you for sharing your experience. Wound management requires medical supervision, with or without honey.

SUGAR

Q My father, who is in hospice care, developed a bedsore on his back. It became infected, and the caregivers used an old wives' remedy of applying sugar to the infected sore. This seemed to work; it had to be discontinued, however, because the agency personnel are not supposed to apply it. I guess the remedy is not approved by the U.S. Food and Drug Administration. Have you ever heard of this remedy? I am now in charge of applying the sugar, and it seems to work.

A It is certainly an old wives' tale, but evidence backs up this approach to treating stubborn wounds. Decades ago, surgeon Richard A. Knutson published this old-fashioned approach to wound care.[2] He resisted his nurse's recommendation at first: "When we started I thought it was absolutely nuts." But his experience with more than 5,000 patients convinced him sugar was useful to speed healing. Recently, scientists compared honey and sugar as wound dressings. Researchers concluded that honey is somewhat more effective than sugar in reducing bacterial contamination and promoting healing.[3]

Q You recently wrote about using sugar for slow-healing wounds and bedsores. As a nurse, I learned years ago that the best way to use this home remedy is to make a thick paste of antibiotic ointment and sugar and to pack the wound with it. Old wives' tale or not, it works. The antibiotic ointment helps to prevent infections.

A We heard from other nurses and even a vet who have not forgotten this old-fashioned treatment. One wrote, "As a nursing student in 1961, I worked at a small hospital that routinely used a mixture of milk of magnesia and sugar to cure bedsores. It seemed to be successful in many cases." Another objected to our terminology: "Using sugar for bedsores is not an old wives' tale. I have been a registered nurse for 45 years. When I was a student, it was very common practice to use sugar packs." The veterinarian said, "Many wounds have been shown to heal three times faster with the use of sugar granules on a saline wet-to-dry bandage. The sugar helps to pull the bacteria from the wound, and the saline feeds the tissue to promote rapid healing of the skin beneath."

Q I want to thank you for the sugar cure column. I had a toe amputated in March 2006, and by November 2007 it still had not healed. I read your column about using sugar for wound healing, took it to my wound treatment appointment, and asked the doctor about it. She said, "Nothing else is helping, so go for it. It couldn't hurt." I applied the first treatment on a Friday afternoon, and by Monday afternoon the improvement was noticeable. At my next appointment, the doctor was very impressed. I mixed the sugar into Polysporin and applied it. The improvement was so great that my doctor even gave me permission to shower without a bag on my foot. I had not done that in well over two years. Healing has slowed, but it continues. Thank you!

A We first found this old-fashioned approach in the medical literature two decades ago. There has been little research on it since then, but research in Africa suggests that both honey and sugar may be helpful.[4] Medical supervision is essential whenever difficult wounds are treated.

EATING FOR HEALTH

THIS SECTION OF THE BOOK DESCRIBES three diets that have proven health benefits: Dietary Approaches to Stop Hypertension (DASH), the Mediterranean diet, and a low-carbohydrate approach. Each of these diets has vocal proponents—as well as some critics—and each has a slightly different emphasis. DASH was designed to lower blood pressure; the Mediterranean diet may decrease the risk of developing heart disease and cancer; and people often adopt a low-carb diet to lose weight or to control blood lipids.

The truth is, however, that these diets are more similar than different, and we're not advocates of one over another. All three involve eating lots of different vegetables, sticking to lean proteins like fish and legumes, and using whole rather than processed grains. And all three seem to offer a wide range of benefits. For instance, research shows that a low-carb diet may help not only with weight loss and blood sugar control but also with serious heartburn symptoms.[1] And in addition to lowering blood pressure, a DASH diet may help to preserve cognitive function.[2]

There are some disparities among the diets, however. For example, the low-carb approach allows you to eat more healthy fats and discourages the consumption of lots of fruit, while DASH emphasizes fruits and vegetables and lets you eat more whole grains. We recommend adopting the diet that makes the most sense for your health considerations and lifestyle. While some of the experts we have cited in this book strongly support one diet over another, we tend to mix and match in our kitchen.

The DASH Diet

The first scientific articles confirming DASH's effectiveness started showing up more than ten years ago.[1] Since then, investigators have published more findings, and today roughly 200 studies are indexed in PubMed.

In the original DASH trial, people with mildly elevated blood pressure ate meals from one of three different eating plans. (None of the volunteers were taking blood pressure medication.) The first group followed a typical American diet; the second ate many more fruits and vegetables (eight to ten servings daily instead of the usual four or fewer); and the third group got eight to ten servings of fruits and vegetables plus low-fat dairy products, so that fat made up only 25 percent of their total calories.

Fruits and vegetables made a difference: Both of the groups with fruits and vegetables in their diet experienced drops in blood pressure readings. Blood pressure fell most in people on the low-fat regimen—the changes in their readings were comparable to those that might be achieved with drugs. In a later twist on the trial, researchers found that cutting people's sodium while sticking with an abundance of veggies and fruits brought blood pressure down even more effectively.[2]

The secret ingredient in fruits and vegetables isn't entirely clear. Some scientists have assumed that it is potassium. For years, studies have suggested that a diet rich in potassium can normalize blood pressure and reduce the risk of stroke.[3] Vegetables are a good source of potassium in the diet, but a recent experiment found that simply supplementing a standard American diet with potassium, fiber, and magnesium did not work as well as putting people on a DASH diet.[4]

Research indicates that a DASH diet can reduce the risk of kidney stones,[5] lower blood pressure, and cut the risk of heart attack and stroke. A low-fat diet with plenty of veggies may also help people lose weight. For some heavy people with mild hypertension, losing ten pounds or so is even better than taking medicine. But nobody can claim that it is easy.

So how can you get in on the benefits of a DASH diet? The basic guidelines are below, but for more information you can go online: *dashdiet.org* or *www.nhlbi.nih.gov/health/public/heart/hbp/dash/new_dash.pdf*. Additional information can be found in *The DASH Diet for Hypertension: Lower Your Blood Pressure in 14 Days—Without Drugs,* written by Thomas Moore and his colleagues, the investigators who ran the original study.

The DASH diet focuses on produce—fruits and vegetables. Most people should eat four servings of vegetables and four servings of fruit every day, with a serving of nuts or beans every other day or so, as well as a couple of servings daily of low-fat dairy products. Two daily servings of lean animal protein are plenty for nonvegetarians. A serving is defined as the size of a deck of cards. It is easier to keep animal protein lean enough by eating fish at least twice a week and by substituting lean ground turkey breast (or even soy crumbles) for fatty ground beef. To get enough calories, several servings of whole grains provide additional daily nutrition. (Here we define a serving as one-half cup of oatmeal or brown rice, or a slice of whole wheat bread.)

One of the biggest challenges with the DASH plan is to keep track of servings of fruits and vegetables to get enough. A daily checklist like this one can be helpful:

Whole Grains • 6 servings daily	☐☐☐☐☐☐
Vegetables • 4 servings daily	☐☐☐☐
Fruits • 4 servings daily	☐☐☐☐
Low-Fat Dairy • 2 servings daily	☐☐
Lean Protein • 2 servings daily	☐☐
Beans/Nuts • ½ serving daily	◿

DASH's sweets allowance is one serving every other day, so it makes sense to use some fruit servings as dessert at the end of lunch or dinner. A single serving of a sweet is a tablespoon of maple syrup, honey, or sugar; one-half cup of frozen yogurt or sherbet; a Popsicle; or a ten-gram piece of dark chocolate. With some practice, you may be able to get by with fewer sweets, which could help with weight loss.

Favorite Vegetables

A serving of vegetables is a cup of salad greens (spinach, romaine lettuce, arugula, and so on). If the vegetables are cooked, go for one-half cup as a serving. Some vegetables are quite expensive when not in season, but once in a while you can find them at a reasonable cost. Here are some of our favorites:

artichoke	carrot	snow pea
arugula	cauliflower	spinach
asparagus	celery	squash (summer or winter)
beet greens	collard greens	sweet potato
beet	green bean	swiss chard
bell pepper	kale	tomato
broccoli	mushroom	V8 juice
cabbage	onion	

Favorite Fruits

Count a medium-size piece of fruit as one serving. You can also substitute six ounces of fruit juice (make sure it is really juice and not sweetened, flavored water), half a grapefruit, or one-half cup berries or cut-up fruit (such as pineapple or mango). If the fruit is dried, the serving size shrinks to one-quarter cup. Here are some of our favorites:

apple	currant (dried)	pineapple
apricot	grapefruit	plum
banana	grape	pomegranate
blackberry	kiwi	raisin
blueberry	orange	raspberry
cantaloupe	peach	strawberry
cherry	pear	tangerine

ADDING FLAVOR

Don't forget the seasonings! Curry, garlic, ginger, lemon, parsley, and salsa don't contribute a lot of bulk to a plate, but they do offer valuable phytonutrients—nutrition from plant sources—along with a boost to flavor. If you are cutting back on salt, a substitute such as one of the Mrs. Dash products can be a great addition to the DASH diet.

The Mediterranean Diet

For years, researchers have been accumulating data on the remarkable health benefits of the Mediterranean diet. Rigorous scientific studies have shown that the diet protects people from many maladies, from diabetes and heart disease[1] to cancer,[2] metabolic syndrome,[3] depression,[4] Alzheimer's disease,[5] and death.[6] Yes, death—or at least the premature kind.

Okay, so the evidence is there. And your intentions may be there, too. But like a lot of other people, perhaps you're wondering: What exactly is the Mediterranean diet?

Broadly, the Mediterranean diet includes lots of vegetables, fruits, and nuts; plenty of legumes (peas, beans, and lentils); fat calories from monounsaturated rather than saturated fats (for example, olive oil rather than butter); moderate consumption of alcohol; and low consumption of meats. For animal protein, fish is preferred to poultry or red meat, and dairy products are consumed somewhat sparingly. Any grains eaten should be whole grains, such as bulgur wheat, barley, or whole wheat bread.

More specifically, there are a few tenets:

- **Eat fish or shellfish at least twice a week.** These foods are highly recommended:

- arctic char (farmed)
- cod from the Pacific Ocean
- halibut from the Pacific Ocean
- herring from the Atlantic
- mahimahi
- pollock from Alaska
- rainbow trout (farmed)
- Alaska salmon (wild or canned)
- sardine (canned)
- tuna (light meat, not albacore; canned)

If you don't like fish—or its high cost—augment your diet with high-quality fish oil supplements from a health food store. You may want to do this even if you are eating two or more servings of fish per week.

- **For protein, substitute at least one serving per day of legumes instead of meats.** Here are some recommended legumes:

- black bean
- chickpea
- kidney bean
- lentil
- navy bean
- pinto bean
- split pea
- soybean

- **Aim for nine or more servings per day of antioxidant-rich vegetables and fruits,** including some of our favorites:

- artichoke
- asparagus
- blackberry
- blueberry
- beet
- broccoli
- brussels sprout
- cherry
- cucumber
- eggplant
- fig
- grape
- kale
- kiwi
- melon
- onion
- orange
- pepper (both bell and hot)
- plum
- pomegranate
- prune
- raspberry
- spinach
- strawberry
- sweet potato
- tomato
- and many more

It shouldn't be hard to find several that you like on this list, and you don't have to eat nine servings of different fruits and veggies per day (although variety is good)—just nine servings total. A serving is typically about one-half cup. With uncooked greens like spinach, a serving is one cup, and with a whole piece of fruit, like an apple, a serving is that piece.

- **Eat roughly one handful per day of tree nuts.** Tree nuts are high in good fats and low in saturated fats. They do have a lot of calories, though, so try to stick to that one handful. And it's best to eat the nuts raw or dry roasted, rather than salted or honey roasted. Toasting them a bit in a skillet makes them tasty. Walnuts are fantastic; other healthful tree nuts include pecans, almonds, and hazelnuts.

- **Stick to monounsaturated and polyunsaturated fats,** like olive oil, canola oil, and oils from nuts—especially walnut oil—and skip the butter.

- **Incorporate whole grain bread, pasta, rice, barley, and couscous into your diet.**

- **The occasional glass of red wine** is fine for most people, but moderation is key. One glass per day with food is okay.

- **Eat lean red meat infrequently**—a few times per month at most.

- **Exercise regularly.** Experts recommend half an hour to an hour five times per week. It makes sense to find something you love to do, whether it is ballroom dancing or walking the dog. That way you will look forward to it and will be more likely to stick with it. Enlisting friends to join you can also help you keep with the program.

- **Eat meals with family and friends.** In traditional Mediterranean countries like France, Italy, and Spain, mealtime is a social gathering. Health and happiness are contagious!

The Low-Carb Diet

Most people can easily get on board with the idea of the Mediterranean diet or the DASH diet. After all, it's hard to argue with data that links good health to consuming fish, vegetables, legumes, whole grains, red wine, and olive oil. The low-carb diet, however, has generated controversy from the day in 1972 when Robert Atkins published his first book, *Dr. Atkins' Diet Revolution.*

Atkins was not the first physician to suggest that sugar and starch caused weight problems. In 1862 a British surgeon, William Harvey, recommended a low-carb diet to an overweight patient named William Banting. Over the next 21 months, Banting lost 50 pounds. Banting wrote up his successful diet in a little pamphlet called "Letter on Corpulence, Addressed to the Public." The pamphlet went on to become wildly successful in Europe and the United States. Not surprisingly, the medical establishment of the day was highly critical. Banting and his followers were advised to mind their own business and not to meddle in medical matters such as obesity.

For more than a century, doctors, diet experts, and medical journal contributors vilified the low-carb concept. A low-fat approach seemed so much more sensible. For one thing, it fits so nicely with our Puritan heritage. We like to suffer, or at least we like to think suffering is good for the soul. Since fatty foods generally taste good (think Brie, sirloin, ice cream, and cheesecake), trying to eliminate them makes us feel holy—or at least superior to the sinners who still consume red meat.

The only trouble with this belief system is that it was built from a house of cards that came tumbling down once there was real data. But old ideas die hard. Too often science takes a backseat to decades of shouting by diet dictocrats. No matter how many studies demonstrate that a low-carb approach seems to help people lose weight and improve blood lipids, the studies tend to disappear without a trace.

On March 7, 2007, the *Journal of the American Medical Association* published an article titled "Comparison of the Atkins, Zone, Ornish, and LEARN Diets for Change in Weight and Related Risk Factors Among Overweight Premenopausal Women: The A TO Z Weight Loss Study: A Randomized Trial."[1] This was a yearlong smackdown—a food fight of epic proportions that pitted the most popular diets of the day against each other. (LEARN represented a standard, physician-recommended low-fat, high-carb, reduced-calorie regimen.)

The results were heretical. The Atkins diet not only helped people shed more pounds than the other diets but also benefited people in other important ways. Their blood fat profiles improved, with higher levels of good high-density lipoprotein (HDL) cholesterol and lower levels of bad triglycerides among the Atkins dieters than among the others. Most dietitians would have predicted that Atkins dieters would develop higher cholesterol levels than their low-fat counterparts. After all, these people were eating red meat, butter, and other dietetically disapproved foods. There were, however, no significant differences in total cholesterol among the groups. Atkins dieters also reduced their blood pressures more effectively than people in the other groups. This may have been partly a result of more weight lost, but even after making that statistical adjustment, Atkins dieters' blood pressures came out lower.

These findings are not as new as they seem. Prior studies have consistently shown that a low-carb diet can lower blood pressure and triglycerides and raise HDL cholesterol.[2] Studies also indicate that people may have better control of their blood sugar on such a program.[3]

If you are open to data rather than dietary dogma, we recommend *Good Calories Bad Calories: Challenging the Conventional Wisdom on Diet, Weight Control and Disease* (Knopf, 2007) by Gary Taubes. It is well researched and fully documented.

Perhaps you are now convinced that a low-carb approach is worth consideration. Which diet is best for you? There are lots to choose from. We have no particular favorites. Whether you select the Atkins approach, the Diabetes Diet (see Richard Bernstein's book), the South Beach Diet, or the Zone, they are all relatively low in carbohydrates. Such an approach is also a low-glycemic diet. That means it is less likely to make blood sugar rise quickly with a concomitant rapid increase in insulin.

So here's the skinny on a low-carb approach: Avoid sugar, starches, and refined carbohydrates. Be especially careful about white foods. Just say NO to foods like these:

- bread
- candy
- crackers
- cookies
- carrots
- cereal
- corn
- fruits and fruit juices
- pancakes
- pasta
- potatoes
- pretzels
- rice
- rye
- sugar
- wheat
- yogurt (sweetened, low-fat)

Here are some of our favorite low-carb foods, based on Richard Bernstein's recommendations from *The Diabetes Diet:*

- asparagus
- beet greens
- bell peppers
- bok choy
- broccoli
- brussels sprouts
- cabbage
- cauliflower
- celery
- cheese
- collard greens
- cottage cheese
- eggplant
- eggs
- endive
- escarole
- fish
- fowl
- green beans
- meat
- mushrooms
- nuts
- okra
- pumpkin
- radicchio
- sauerkraut
- scallions
- seafood
- snow peas
- spaghetti squash
- spinach
- tofu
- yogurt (full-fat)
- zucchini

As with any diet, regular exercise—half an hour to an hour five times per week—is essential for maintaining weight and improving overall health. And whatever diet you choose, be sure to take time to enjoy your food with family and friends.

BREAKFAST

Anti-inflammatory Curcumin Scramble

4 large eggs (if you prefer, substitute 2 egg whites for 1 egg)
Pinch salt and ground black pepper
1 teaspoon turmeric (curcumin) powder
1 tablespoon olive or grapeseed oil
2 tablespoons milk or water (milk will make the eggs denser and creamier; water will make them fluffier and lighter)
½ teaspoon grated Parmesan
Sprinkling chopped chives

In a medium-size mixing bowl, use a fork to mix eggs, salt, pepper, and curcumin powder. (Be very careful with the curcumin powder; it will dye everything it touches a bright, summery yellow.) Add olive or grapeseed oil to a large sauté pan or skillet, and set your burner to medium-low heat. While the oil is heating, stir milk or water into your egg mixture. When oil is hot, pour egg mixture into pan or skillet. Let eggs set for about 20 seconds before gently scrambling with a wooden spatula. Add Parmesan. Let set for another 15 seconds, and gently scramble again. Repeat until eggs look almost but not quite cooked. (They'll continue to cook on the plate.) Serve, and add a sprinkling of chives to each dish for garnish. Makes two servings.

Joe's Brain-Boosting Smoothie

1 teaspoon fish oil
1 frozen banana
1 cup fresh or frozen mixed berries
If using fresh berries: 1 cup crushed ice
¾ cup plain yogurt
½ cup pomegranate or cherry juice
½ cup pasteurized egg whites (roughly 4 eggs' worth)
4 tablespoons whey powder
2 tablespoons ground flaxseed

Add all ingredients to a blender. Puree until smooth. You may need to scrape down the sides once or twice with a spatula to ensure that the mixture is evenly blended. Pour into tall glasses and enjoy right away, or chill in the refrigerator overnight. Makes two generous servings.

Cholesterol-Combating Oatmeal

1 cup soy or rice milk
¼ cup steel-cut oats
1 teaspoon ground cinnamon
Pinch salt
½ cup egg whites
¼ cup dried fruit (we like currants, raisins, and cherries), or ⅔ cup fresh fruit
2 tablespoons ground flaxseed
⅓ cup chopped walnuts or almonds
2 teaspoons honey, agave nectar, maple syrup, or sugar substitute (optional)

Bring soy or rice milk to a boil in a medium saucepan. (We recommend using a lid so that no milk escapes as it begins to boil.) Add oats, cinnamon, and salt. Stir with a slotted spoon and reduce heat to a simmer. Continue stirring while adding the egg whites, and then cook uncovered over low heat for about ten minutes while stirring occasionally. Add fruit, flaxseed, and nuts, and cook until oats are tender, approximately five to ten minutes more. Top each serving with sweetener to taste, if desired. Makes two generous helpings.

LUNCH

Crustless Cauliflower and Red Pepper Quiche

Contributed by Eric C. Westman, M.D., M.H.S.

1 teaspoon olive oil, plus more for pie plate
½ small onion, finely chopped
½ red bell pepper, diced
4 large eggs
½ cup heavy cream
1 cup water
1 cup grated Monterey Jack cheese, divided
¼ teaspoon dried thyme
¼ teaspoon dried oregano
¼ teaspoon chopped dried rosemary
½ teaspoon salt
¼ teaspoon pepper
1 small head cauliflower, cut into florets, stems peeled
  and cut ⅓ inch thick (substitute broccoli, if you prefer)

Heat oven to 375 degrees. Brush a nine- or ten-inch pie plate with olive oil. Heat oil in a small skillet over medium-high heat. Add onion and red pepper and cook until softened, about three minutes. Transfer to a medium bowl and let cool. Add eggs to onion mixture and lightly beat. Whisk in cream, water, ½ cup cheese, herbs, salt, and pepper to blend. Cover bottom of pie plate with cauliflower. Cover with cream mixture and sprinkle with remaining ½ cup cheese. Bake until a knife inserted in the middle comes out clean and quiche is golden brown, 50 to 60 minutes. Makes six servings.

Sole with Lemongrass Red Bell Pepper Salad

Contributed by Mark Liponis, M.D.

3 red bell peppers, roasted and thinly sliced
3 tablespoons minced lemongrass
1½ tablespoons sherry vinegar
¾ teaspoon salt
¼ teaspoon pepper
4 sole fillets, about 4 ounces each
¼ teaspoon salt
¼ teaspoon pepper
1 tablespoon olive oil

In a medium bowl, combine the first five ingredients and mix well. Set aside. Season fish with salt and pepper. Sauté in a large sauté pan in olive oil about three to five minutes on each side or until cooked through. Serve one fish fillet with ⅓ cup of the salad. Makes four servings.

Wheat Berry Salad

Contributed by Jeffrey Blumberg, Ph.D., and Helen Rasmussen, Ph.D., R.D., F.A.D.A.

1 cup prepared wheat berries (see preparation method, below)
½ teaspoon chicken bouillon (cube or loose granules; can use low-sodium bouillon)
⅓ cup water
2 tablespoons lemon juice
½ tablespoon olive oil
2 tablespoons finely sliced onion
⅛ teaspoon ground black pepper
1 clove garlic, minced or pressed
1 tablespoon dry wheat bran
2 tablespoons chopped walnuts

TO PREPARE WHEAT BERRIES: Put one cup dry wheat berries into a saucepan (or use a rice cooker, if you prefer). Add 2½ cups water and bring to a boil. Stir the berries, lower the heat to a simmer, cover, and cook for 20 minutes. Regularly check to see that there is still adequate water to prevent scorching. Cook for another 40 minutes, or until the berries are soft.

TO PREPARE SALAD: Dissolve the chicken bouillon in ⅓ cup water. In a serving bowl, mix the dissolved bouillon, lemon juice, and olive oil. Stir in onion, pepper, and garlic. Add the cup of cooked wheat berries, wheat bran, and walnuts. Mix together. Makes one serving.

DINNER

Cardamom Grilled Chicken with Mango Lime Sauce

Contributed by Mark Liponis, M.D.

Spice mix:
3 tablespoons cardamom
1 tablespoon black pepper
1 tablespoon salt
1 teaspoon cinnamon
¼ teaspoon cayenne pepper
Six 4-ounce skinless chicken breasts, boned and defatted

Sauce:
1 mango, cleaned and diced
½ cup lime juice
2 tablespoons olive oil
1 tablespoon minced ginger
½ cup nonfat plain yogurt
½ teaspoon salt
1 tablespoon diced jalapeño
1 tablespoon chopped cilantro

Prepare coals for grilling or preheat broiler. In a small bowl, combine ingredients for spice mix. Lightly pound chicken breasts to flatten. Dust each with about one teaspoon spice mix. Store remaining spice mix in an airtight container for future use. Grill chicken for three to five minutes on each side, or until juices run clear when pierced with a fork. Combine all ingredients for sauce in a blender, except for jalapeño and cilantro. Puree until smooth. Pour into a bowl, add jalapeño and cilantro, and gently stir. Serve one chicken breast with one-quarter cup sauce. Makes six servings.

Lentil Nut Loaf with Red Pepper Sauce

Contributed by Gail Pettiford Willett and Walter C. Willett, M.D., Dr.P.H.

Lentil nut loaf:

1 cup lentils, washed
2 cups water
2 tablespoons olive oil, plus additional oil for greasing the baking pan
1 large onion, chopped
1 cup mushrooms, chopped
1 cup walnuts, chopped
1 cup whole wheat breadcrumbs
1 tablespoon lemon juice
1 tablespoon soy sauce
Salt and pepper to taste
1 tablespoon mixed herbs of your choice

Red pepper sauce:

2 tablespoons olive oil
3 cloves garlic, peeled and chopped
1 small jar roasted red peppers (4-6 oz.)
3 tomatoes, chopped
Red pepper flakes (optional)

FOR LENTIL NUT LOAF: Preheat oven to 350 degrees. Grease a baking pan. Combine lentils and water in a large pot, and cook lentils until they're soft, about half an hour. Heat the olive oil in a sauté pan or skillet, and sauté the onion and mushrooms until soft. Mix all other ingredients in with the onion and mushrooms. Sauté for three to four minutes. Place the mixture in the greased baking pan and bake for 30 minutes.

FOR RED PEPPER SAUCE: Heat oil in sauté pan or skillet, and sauté the garlic on medium heat for three to four minutes. Add the red peppers and tomatoes. Cook until the mixture thickens, about 15 minutes. Ladle over the lentil nut loaf. Can be served hot or at room temperature. Makes four servings. (Lentil loaf recipe adapted from Recipes for Natural Health: *www.recipes.org/health/main.htm;* red pepper sauce recipe adapted from *yumyum.com.*)

Pescado al Cilantro

This recipe is of a Mexican style, but simpler. The sauce is spicy, and the jalapeños are salty, so there's no need to add salt during preparation.

⅔ cup cilantro leaves, rinsed and destemmed
2 cloves garlic, peeled and chopped
6 tablespoons pickled jalapeño peppers, with the juice
4 filets of firm, white-fleshed fish, such as flounder, tilapia, cod, or haddock
Black pepper to taste
Additional cilantro leaves and lemon wedges for garnish

Preheat the broiler. Put the cilantro leaves, garlic, and pickled jalapeños with their juice into a blender. Process mixture into a paste.

Place the fish onto a broiler pan and spread the blended sauce over the filets. Grind black pepper over them. Broil five to ten minutes, depending on the thickness of the filets, until just cooked through. Sprinkle additional cilantro leaves over cooked filets for garnish, and serve with cut lemon wedges. Makes four servings.

Coleslaw with Mint

Coleslaw:

½ head red cabbage, chopped finely
¼ cup onion (red or sweet), chopped finely
1 green bell pepper, deribbed and chopped
1 tablespoon fresh mint leaves, chopped finely

Dressing:

2 tablespoons Dijon mustard
¼ teaspoon salt and freshly ground pepper to taste
4 drops Tabasco sauce
⅓ cup apple cider vinegar
¼ cup olive oil

Mix together all ingredients for the coleslaw and set aside. To make the dressing: Whisk ingredients together in a bowl, or pour into a large jar, close the lid tightly, and shake until well blended. Pour the dressing over the slaw and mix well. Makes four servings.

Farmers Market Saag

3 bunches fresh greens (mix and match, depending on what's in season: collards; kale; chard; beet, mustard, or turnip greens; spinach; arugula)
3 or 4 turnips or beets, or 2 sweet potatoes
1½ tablespoons safflower or grapeseed oil
1 teaspoon cumin seeds
2 cloves garlic, pressed or finely minced
1 fresh jalapeño, seeded and minced (optional—but good)
1 teaspoon finely minced ginger, or ½ teaspoon ginger powder
1 teaspoon kosher salt
1¼ cups boiling water
1 teaspoon garam masala

Wash greens thoroughly and chop coarsely. Scrub and trim turnips or beets or pare sweet potatoes, and cut into two-inch pieces. Heat the oil in a large frying pan, preferably cast iron. When the oil is hot, add cumin seeds and stir until they darken (just a few seconds). Then stir in garlic and jalapeño (if desired). Add root vegetables and sauté for six to ten minutes, regulating the heat so they don't burn. Stir in chopped greens by the handful, adding more as they wilt. Sprinkle them with ginger and salt, and stir well. Add the boiling water, reduce the heat, cover the pan, and cook for about 20 minutes, or until the root vegetables are tender. Uncover and cook a bit more, to reduce excess liquid. Stir in the garam masala and remove the saag from heat. Makes four servings.

Roasted Garlic

Contributed by Michael D. Ozner, M.D., F.A.C.C., F.A.H.A.

1 jumbo garlic head
1 hearty pinch garlic powder or garlic salt (optional)
Olive oil to drizzle

Heat oven to 400 degrees. Remove only loose leaves from the garlic bulb. Holding the head firmly with stem up, cut off the pointy tops of each clove. Try to keep the remaining leaves and cloves intact. Place head (stem and cut cloves facing up) in a tight-fitting baking dish. Sprinkle the top of the head with garlic powder or garlic salt, if desired, and drizzle with olive oil. Place in the oven and bake until the head is golden brown on top and the clove tops can be easily pierced with a fork, about 20 to 30 minutes.

Simple Salad Dressing

Contributed by Helen Graedon

½ cup red wine vinegar
2 or 3 tablespoons olive oil
At least 2 cloves fresh garlic, minced or pressed
2 teaspoons Dijon mustard
Chopped green onions (optional)
Salt and freshly ground pepper to taste

Blend all ingredients. One of the best salad dressings around.

SOURCES

Notes, Introduction

1. *Integrative and Comparative Biology* 2009
2. *Molecular Cancer Therapeutics* 2010; *Osteoarthritis Cartilage* 2010; *Alternative Medicine Review* 2009; *Circulation Journal* 2010; *Journal of Investigative Dermatology* 2010
3. *Hypertension* 2010
4. *American Journal of Clinical Nutrition* 2001; *American Journal of Clinical Nutrition* 2009; *American Journal of Clinical Nutrition* 2009
5. *Journal of the American Medical Association* 2002
6. *Journal of Nutrition* 2009

Notes, Home Remedies

ACNE AND ROSACEA

1. *Journal of Investigative Dermatology* 2010
2. *Archives of Dermatology* 2002
3. *International Journal of Dermatology* 1995
4. *Cutis* 2006
5. *Archives of Dermatology* 1975

ALLERGIES

1. *Wisconsin Medical Journal* 2008
2. *Bioscience, Biotechnology, and Biochemistry* 2001
3. *Alternative Medicine Review* 2007

ASTHMA

1. *Cochrane Database of Systematic Reviews* 2003

BRAIN FUNCTION

1. *Journal of Agricultural and Food Chemistry* 2008

BURSITIS

1. *Clinical and Experimental Rheumatology* 2006

COLDS

1. *Chest* 2000
2. *Family Practice* 2005
3. *Pediatrics* 2009

COLD SORES

1. *Alternative Medicine Review* 2006

CUTS AND BRUISES

1. *Archives of Facial Plastic Surgery* 2006

DIABETES

1. *Journal of Medicinal Food* 2010
2. *Diabetes Care* 2006
3. *European Journal of Clinical Nutrition* 2005
4. *Diabetes Care* 2004

DIARRHEA

1. *Digestive Diseases and Sciences* 2007

ECZEMA

1. *British Medical Journal* 2003
2. *Archives of Dermatology* 2001
3. *Journal of Allergy and Clinical Immunology* 2008

GOUT

1. *Journal of Natural Products* 2006
2. *Journal of Nutrition* 2003
3. *Lancet* 2004
4. *New England Journal of Medicine* 2004

HEADACHES AND MIGRAINES

1. *Journal of Ethnopharmacology* 1990
2. *Medical Science Monitor* 2005

HEARTBURN

1. *The Lancet* 1990
2. *Digestive Diseases and Sciences* 2004
3. *New England Journal of Medicine* 1984
4. *New England Journal of Medicine* 2002
5. *Alternative Therapies in Health and Medicine* 2001
6. *Clinical Gastroenterology and Hepatology* 2009
7. *Digestive Diseases and Sciences* 2006
8. *Journal of the American Medical Association* 2004, 2005

HIGH BLOOD PRESSURE

1. *Hypertension* 2005; *Archives of Internal Medicine* 2006
2. *Archives of Internal Medicine* 2008
3. *The Lancet* 2005
4. *Cardiovascular Disorders* 2008
5. *Hypertension Research* 2009
6. *American Journal of Clinical Nutrition* 2004
7. *Archives of Internal Medicine* 2004
8. *Diabetes Care* 2004
9. *Journal of Human Hypertension* 2005
10. *British Medical Journal* 2007
11. *Food Chemistry and Toxicology* 2008
12. *Journal of Clinical Investigation* 2008

HIGH CHOLESTEROL AND TRIGLYCERIDES

1. *American Journal of Cardiology* 2008
2. *British Journal of Nutrition* 2006

HOT FLASHES

1. *Annals of Internal Medicine* 2006
2. *Photochemistry and Photobiology* 2007
3. *Acta Obstetricia et Gynecologica Scandinavica* 2007
4. *Acta Obstetricia et Gynecologica Scandinavica* 2007
5. *Acta Obstetricia et Gynecologica Scandinavica* 2007
6. *Journal of the American College of Nutrition* 2005

IRRITABLE BOWEL SYNDROME

1. *Journal of Gastroenterology* 1997
2. *Phytomedicine* 2005

JOINT PAIN AND ARTHRITIS

1. *Journal of Nutrition* 2005
2. *Scandinavian Journal of Rheumatology* 2006
3. *Journal of Nutrition* 2003
4. *Pain* 2007
5. *Complementary Therapies in Medicine* 2009
6. *Clinical and Experimental Rheumatology* 2006

KIDNEY STONES

1. *Urology* 2007
2. *British Journal of Urology International* 2003
3. *Annals of Internal Medicine* 1998
4. *Journal of Urology* 2001
5. *Journal of Urology* 2007

LICE

1. *Pediatrics* 2004

MACULAR DEGENERATION

1. *Critical Reviews in Food Science and Nutrition* 2009; *Ophthalmology* 2008
2. *Ophthalmology* 2009
3. *Biofactors* 2008
4. *Archives of Ophthalmology* 2001
5. *Eye* 2008
6. *Archives of Ophthalmology* 2009
7. *Archives of Ophthalmology* 2007
8. *American Journal of Clinical Nutrition* 2007
9. *Archives of Ophthalmology* 2001

MOTION SICKNESS, VERTIGO, AND DIZZINESS

1. *Harvard Women's Health Watch* 2008
2. *Health Notes* 1999

MUSCLE AND LEG CRAMPS

1. *The Lancet* 2002

NAUSEA

1. *Cochrane Database of Systematic Reviews* 2009

NERVE PAIN

1. *Progress in Neurobiology* 2009
2. *European Journal of Pharmacology* 2009
3. *Endocrine, Metabolic & Immune Disorders—Drug Targets* 2009
4. *American Journal of Human Genetics* 2007

PLANTAR FASCIITIS

1. *American Journal of Veterinary Research* 2009; *British Journal of Sports Medicine* 2006
2. *Pain* 2007

PSORIASIS

1. *Advances in Experimental Medicine and Biology* 2007
2. *Annals of the New York Academy of Sciences* 2004
3. *Biochemical Pharmacology* 2007
4. *British Journal of Dermatology* 2000
5. *American Journal of Clinical Nutrition* 2008

RESTLESS LEG SYNDROME

1. *Geriatrics* 2007
2. *Movement Disorders* 2004
3. *Sleep* 1998

SEX/LIBIDO

1. *Journal of Sex and Marital Therapy* 2003
2. *Menopause* 2006
3. *Blood Pressure* 2003
4. *Asian Journal of Andrology* 2007

SUNBURN AND SUN RASH

1. *Journal of Investigative Dermatology* 2005

WARTS

1. *Annals of Pharmacotherapy* 2007
2. *Archives of Pediatric and Adolescent Medicine* 2002
3. *Archives of Pediatric and Adolescent Medicine* 2006
4. *Archives of Dermatology* 2007
5. *Virology* 2008

WEIGHT LOSS

1. *American Journal of Clinical Nutrition* 2004

WOUND CARE

1. *Diabetes Research and Clinical Practice* 2010
2. *Southern Medical Journal* 1981
3. *Journal of Wound Care* 2007
4. *Journal of Wound Care* 2007

Notes, Favorite Foods

COFFEE

1. *American Journal of Obstetrics and Gynecology* 2008
2. *American Journal of Obstetrics and Gynecology* 2009
3. *Nutrition Reviews* 2007
4. *American Journal of Epidemiology* 2008
5. *International Journal of Cancer* 2009
6. *International Journal of Cancer* 2007
7. *International Journal of Cancer* 2007
8. *Journal of Investigative Dermatology* 2009; *Journal of Investigative Dermatology* 2009
9. *International Journal of Cancer* 2009
10. *Circulation* 2009
11. *Journal of Alzheimer's Disease* 2009

BLUEBERRIES

1. *Journal of Nutrition* 2009
2. *Subcellular Biochemistry* 2007; *Journal of Agricultural and Food Chemistry* 2008

See also *Journal of Agricultural and Food Chemistry* 2010

3. *Nutrition Research* 2009
4. *Rejuvenation Research* 2008
5. *Antioxidants & Redox Signaling* 2009
6. *Cancer Research* 2010

GREEN TEA

1. *Genes & Nutrition* 2009
2. *Journal of Nutrition* 2009
3. *BMC Musculoskeletal Disorders* 2009
4. *Cochrane Database of Systematic Reviews* (online) 2009
5. *Blood* 2009

POMEGRANATE

1. *Nutritional Biochemistry* 2005
2. *Journal of Medicinal Food* 2009
3. *Nutrition Reviews* 2009
4. *Journal of Agricultural and Food Chemistry* 2009; *Zhongguo Zhong yao za zhi* 2009
5. *Alternative Medicine Review* 2008
6. *Pakistan Journal of Pharmaceutical Science* 2009

CHICKEN SOUP

1. *Current Infectious Disease Reports* 2002
2. *Chest* 2000
3. *The Lancet* 2006
4. *Current Medicinal Chemistry* 2004

PINEAPPLE

1. *Clinical Immunology* 2008
2. *Cancer Letters* 2009
3. *QJM: Monthly Journal of the Association of Physicians* 2006
4. *Cancer Letters* 2009

CINNAMON

1. *Journal of Agricultural and Food Chemistry* 2004
2. *Diabetes Care* 2003
3. *Phytotherapy Research* 2005
4. *European Journal of Applied Physiology* 2009

YOGURT

1. *International Journal of Colorectal Disease* 2007
2. *Annals of Internal Medicine* 1992

OOLONG TEA

1. *Archives of Internal Medicine* 2004
2. *Stroke* 2009
3. *Archives of Dermatology* 2001

ALMONDS

1. *Archives of Internal Medicine* 2010
2. *American Journal of Clinical Nutrition* 2005
3. *American Journal of Clinical Nutrition* 2008
4. *Metabolism* 2007

BROCCOLI

1. *Journal of Digestive Diseases* 2008
2. *Planta Medica* 2008
3. *Carcinogenesis* 2006
4. *Cancer Letters* 2008
5. *Environmental and Molecular Mutagenesis* 2009

HOT PEPPERS

1. *Archives of Internal Medicine* 2006
2. *New England Journal of Medicine* 2002
3. *Journal of Physiology Paris* 2001
4. *Journal of Physiology Paris* 1997
5. *Gastroenterology* 1989
6. *Inflammopharmacology* 2007
7. *Prostaglandins Leukotrienes & Essential Fatty Acids* 2009
8. *Planta Medica* 2008; *Biochemical and Biophysical Research Communications* 2009; *Cellular and Molecular Biology Letters* 2009

BEETS

1. *Hypertension* 2008
2. *Lancet Oncology* 2010

3. *Hypertension* 2010
4. *American Journal of Physiology—Heart and Circulatory Physiology* 2009

CHOCOLATE

1. *Journal of Nutrition* 2008
2. *Journal of Nutrition* 2009
3. *American Journal of Clinical Nutrition* 2010
4. *Circulation* 2009
5. *Journal of Internal Medicine* 2008

GRAPE JUICE

1. *Journal of Nutrition* 2009
2. *Biofactors* 2004
3. *Journal of Nutrition* 2009
4. *Journal of Nutrition* 2009

FISH AND FISH OIL

1. *New England Journal of Medicine* 2002; *Journal of the American Medical Association* 2002; *Circulation* 2002
2. *New England Journal of Medicine* 2010
3. *Seminars in Neurology* 2006
4. *American Journal of Clinical Nutrition* 2009
5. *Archives of Ophthalmology* 2009

WALNUTS

1. *Archives of Internal Medicine* 2002
2. *Journal of the American Medical Association* 2002

CHERRIES

1. *Scandinavian Journal of Rheumatology* 2006; *Journal of Nutrition* 2006; *Journal of Agricultural and Food Chemistry* 2009
2. *Phytomedicine* 2001
3. *Behavioural Brain Research* 2004
4. *Journal of Nutrition* 2003

MUSTARD

1. *Cancer Letters* 2003

GINGER

1. *American Journal of Physiology—Gastrointestinal and Liver Physiology* 2003
2. *Pharmacology* 1991
3. *Phytomedicine* 2005
4. *Supportive Cancer Therapy* 2007

CURRY

1. *Trends in Pharmacological Sciences* 2009
2. *Molecular Oncology* 2008
3. *European Neuropsychopharmacology* 2009; *Behavioural Brain Research* 2009; *Current Opinion in Pharmacology* 2009; *Current Alzheimer Research* 2009

GARLIC

1. *Journal of Agricultural and Food Chemistry* 2009
2. *Critical Reviews in Food Sciences & Nutrition* 2009
3. *Journal of Nutrition* 2001

Notes, Eating for Health

INTRODUCTION

1. *Digestive Diseases and Sciences* 2006
2. Cache County Study on Memory, *Health and Aging* 2009

THE DASH DIET

1. *New England Journal of Medicine* 1997
2. *New England Journal of Medicine* 2001
3. *Nutrition Reviews* 1999
4. *Journal of Human Hypertension* 2009
5. *Journal of the American Society of Nephrology* 2009

THE MEDITERRANEAN DIET

1. *American Journal of Clinical Nutrition* 2008
2. *British Journal of Cancer* 2008; *Nutrition Reviews* 2009
3. *Public Health Nutrition* 2009
4. *Archives of General Psychiatry* 2009
5. *Journal of the American Medical Association* 2009; *Journal of the American Medical Association* 2009
6. *Journal of the American Medical Association* 2004; *Maturitas* 2009; *British Medical Journal* 2008

THE LOW-CARB DIET

1. *Journal of the American Medical Association* 2007
2. *Journal of the American Medical Association* 2005; *New England Journal of Medicine* 2003; *Mayo Clinic Proceedings* 2003; *International Journal of Cardiology* 2006; *Journal of Nutrition* 2006; *Annals of Internal Medicine* 2004
3. *Metabolic Syndome and Related Disorders* 2003; *Nutrition and Metabolism (London)* 2008

ABOUT THE AUTHORS

Joe Graedon, M.S.

Joe Graedon received his B.S. from Pennsylvania State University in 1967 and did research on mental illness, sleep, and basic brain physiology at the New Jersey Neuropsychiatric Institute in Princeton. In 1971 he earned his M.S. in pharmacology from the University of Michigan. Joe was conferred the degree of Doctor of Humane Letters *honoris causa* from Long Island University in 2006 as one of the country's leading drug experts for the consumer.

Joe taught pharmacology at the School of Medicine of the Universidad Autónoma "Benito Juárez" de Oaxaca, Mexico, from 1972 to 1974. He has lectured at Duke University; the University of California, San Francisco; and the University of North Carolina, where he has been an adjunct assistant professor since 1986. He is a member of the American Association for the Advancement of Science (AAAS), the Society for Neuroscience, and the New York Academy of Sciences. In 2005 Joe was elected to the rank of AAAS fellow for "exceptional contribution to the communication of the rational use of pharmaceutical products and an understanding of health issues to the public."

Joe has served as an editorial adviser to *Men's Health Newsletter* and *Prevention* magazine. He is an advisory board member of the American Botanical Council (*HerbalGram*), and he has served as a member of the Board of Visitors, UNC Eshelman School of Pharmacy, since 1989.

Teresa Graedon, Ph.D.

Medical anthropologist Teresa Graedon received her Ph.D. from the University of Michigan in 1976. Her doctoral research was on health and nutritional status in a migrant community in Oaxaca, Mexico. Teresa taught at the Duke University School of Nursing with an adjunct appointment in the Department of Anthropology from 1975 to 1979 and pursued postdoctoral training in medical anthropology at the University of California, San Francisco. She is a fellow of the Society for Applied Anthropology and a member of the American Anthropological Association and the Society for Medical Anthropology. She served on the Foundation Board of the University of North Carolina School of Nursing and currently serves as a trustee for Guilford College.

Joe & Terry & The People's Pharmacy

Joe and Terry write a newspaper column, "The People's Pharmacy," which has been syndicated nationally by King Features Syndicate since 1978. The *People's Pharmacy* radio show won a Silver Award from the Corporation for Public Broadcasting in 1992. It is syndicated to more than 100 radio stations in the United States, primarily on public radio. Terry and Joe were presented with the *America Talks Health* Health Headliner of 1998 Award for superior contribution to the advancement of medicine and public health education. In 2003 Joe and Teresa received the Alvarez Award at the 63rd annual conference of the American Medical Writers Association for excellence in medical communications. You can communicate with the Graedons through their website, *www.peoplespharmacy.com*

ACKNOWLEDGMENTS

ALENA GRAEDON made this book possible by coordinating, collaborating, and organizing.

DAVE GRAEDON got the ball rolling and helped us collect many of the home remedies in this book.

CHARLOTTE SHEEDY helped make the connections that brought this book to print. She is an amazing agent.

THE RESEARCHERS investigating the evidence for using food as medicine provided the scientific foundation.

THE EXPERT GUESTS on our syndicated radio show shared their wisdom and emphasized the importance of the healing power of food. Those who contributed recipes to this book are listed separately on the next page. We are grateful for their support.

RECIPE CONTRIBUTORS

Jeffrey Blumberg, Ph.D., is a professor in the Friedman School of Nutrition Science and Policy and director of the Antioxidants Research Laboratory at the Jean Mayer USDA Human Nutrition Research Center on Aging at Tufts University.

Helen Graedon was a schoolteacher, political activist for peace and justice issues, and dedicated home cook. She was Joe Graedon's mother.

Mark Liponis, M.D., a practicing clinician in internal medicine for 20 years, is the author of *Ultraprevention* (2005) and *UltraLongevity* (2007).

Tieraona Low Dog, M.D., is clinical associate professor at the Arizona Center for Integrative Medicine. She served on the White House Commission on Complementary and Alternative Medicine and received the 2010 People's Pharmacy Award.

David Mathis, M.D., F.A.A.F.P., A.B.H.M., D.Ay., was the first board-certified family physician to combine ayurveda and family medicine. He is associate professor of primary care medicine at George Washington University.

Debbie Mathis, M.A., D.Ay., studied ayurvedic medicine in Boston, Albuquerque, and India and joins her husband in their pioneering Integrative Medicine practice. She also established the website *Ayurveda-MD.com.*

Sally Fallon Morell is founding president of the Weston A. Price Foundation and author of the best-selling *Nourishing Traditions: The Cookbook That Challenges Politically Correct Nutrition* and *Diet Dictocrats.*

Michael D. Ozner, M.D., F.A.C.C., F.A.H.A., is a leading advocate for heart disease prevention. Dr. Oz, as he is known to many, is a board-certified cardiologist and author of *The Miami Mediterranean Diet* and *The Great American Heart Hoax*. For more information, go to *www.drozner.com.*

Helen Rasmussen, Ph.D., R.D., F.A.D.A., is the senior clinical research dietitian at the Jean Mayer USDA Human Nutrition Research Center on Aging at Tufts University and works with Community Servings, an organization serving home-delivered meals to the critically ill.

Eric C. Westman, M.D., M.H.S., is an associate professor of medicine at Duke University Health System and director of the Duke Lifestyle Medicine Clinic. With Dr. Stephen Phinney and Dr. Jeff Volek, he is the author of *The New Atkins for a New You.*

Gail Pettiford Willett has worked in human services for 30 years. She founded Savanna Books, a multicultural children's bookstore.

Walter C. Willett, M.D., Dr.P.H., is professor of epidemiology and nutrition at the School of Public Health and professor of medicine at Harvard. He is the author of three books including *The Fertility Diet*, with Jorge Chavarro and Pat Skerrett.

INDEX

A

Acne 15–19, 146, 206
Acupressure
  for dizziness 167
  for insomnia 143
  for nausea 185
Age-related macular degeneration 131, 163–166
Allergies 20–21, 93, 142, 209
Almonds
  as favorite food #12 105
  for heartburn 104, 106
  for high cholesterol 105, 106, 135
Aloe vera
  for burns 37, 43
  for canker sores 43–44
Anemia 22–23
Angostura bitters 95, 110
Anthocyanins 152, 153, 192
Antiperspirant
  for skin fungus 210
  for stinky feet 214
Apitherapy 149, 207–208
Arnica gel, for bruises 66–67, 68
Arthritis remedies 148–158
  bee venom therapy 149, 207–208
  Certo and grape juice 129, 133, 150–151, 157
  cherries and cherry juice 94, 100, 151–153, 192
  fish oil 131, 155–155
  gin-soaked raisins 155, 157, 201
  turmeric 11, 99, 128, 152, 154, 198
Aspartame 102, 154
Asthma 23–24
Astragalus root 52, 199, 200
Atkins Diet 113, 226, 238–239
Ayurvedic medicine 21, 54, 94, 98, 189, 197

B

Back pain 24, 26–27, 150, 152, 192
Bacon fat, for warts 220–221
Bag Balm 86, 87
Baking soda
  for body odor 27
  for bug bites and stings 32, 33, 36
  for canker sores 44
  for heartburn 106
  for stinky feet 214–215
Bananas
  acne and 17
  for canker sores 44
  for heartburn 106
  peel, for warts 221
  potassium content 175
Barley, for high cholesterol 130
Barnyard beauty aids, for dry skin 86–87
Bee venom therapy
  for arthritis 149
  for shingles 207–208
Beer
  gout and 100
  for migraines 101
Beet juice, for high blood pressure 121, 122
Beets
  anticancer potential 43
  blood pressure and 11, 121, 122
  cholesterol and 122, 134
  as favorite food #15 122
  iron content 22
  nitrate content 121, 122
Bergamot 170
Bilberry, for macular degeneration 164–165
Bitter melon, for diabetes 73, 75
Bitters, for gas 95, 110
Black cohosh, for hot flashes 137, 138, 139
Black pepper, for cuts 67, 68, 69
Black tea 41, 126
Blackstrap molasses
  for anemia 22–23
  for constipation 60
  for hemorrhoids 116–117
  for muscle and leg cramps 170
Blueberries
  for brain function 11, 30, 31
  as favorite food #2 31
Body odor 27–29
Borage oil
  for eczema 88, 91
  for hot flashes 138
Boswellia
  anti-inflammatory properties 94, 154, 189
  for fibromyalgia 94
  for joint pain and arthritis 149–150, 154
  turmeric and 187, 189
Brain function 11, 30–32, 133, 142, 231
Broccoli
  anticancer potential 43, 110
  as favorite food #13 110
  gas from 110
  for heartburn 107, 109

Bromelain
for bursitis 39–40
for joint pain and arthritis 57, 154, 157
medicinal effects 40, 57, 154, 157, 189
turmeric and 187, 189
Bruises. *See* Cuts and bruises
Bug bites and stings 32–36
Burns 36–39, 43
Bursitis 39–40, 128
Buttermilk
for canker sores 44–45
for cold sores 55

C

Caffeine
addiction 25, 101
for asthma 24
blood pressure and 126
blood sugar and 126–127
iron levels and 22, 23
as migraine trigger 102
as reflux trigger 141
Calcium-rich foods, kidney stones and 159–160
Cancer 40–43
Canker sores 43–48, 55
Castor oil
for ant bites 32–33
for bruises 68
for muscle and leg cramps 171
for warts 33, 68, 221, 224
Cayenne pepper
for arthritis 150
for cuts 68
Celery seed extract, for gout 98
CeraVe moisturizer 85, 89
Certo (plant pectin) and juice
for high cholesterol 129
for joint pain and arthritis 129, 133, 150–151, 157
Cetaphil 85, 161
Cherries
as favorite food #20 153
for fibromyalgia 94, 153
sour, for gout 99–100, 152
Cherry juice
for arthritis 100, 151–152
for plantar fasciitis 192–193
Chewing gum
for heartburn 107–108
sugar-free, diarrhea and 63, 80, 81
Chicken soup
for colds 48–49, 52
as favorite food #5 52
recipes 50, 51
Chinese medicine 115, 132, 196, 199
Chlorophyll tablets 215
Chocolate
arginine content 56
cardiovascular benefits 124
as favorite food #16 124
as heartburn trigger 109, 112
for high blood pressure 123, 125
as migraine trigger 102
as reflux trigger 141
Chocolate chips, for hiccups 117–118
Cinnamon
coumarin in 73, 74, 200
for diabetes 73–75
as favorite food #8 74
for high cholesterol and triglycerides 129–130
for Raynaud's disease 199–200
Cinnamon-ginger drink 109, 111
Coconut
for colitis 59
constipation and 58, 79
for diarrhea 58, 78–80, 145
as favorite food #7 58
for irritable bowel syndrome 145
Coconut macaroon cookies 58, 59, 78–80, 82, 145
Coconut oil
for lice 162
for vaginal dryness 219
Coconut water 58
Coffee
for asthma 23, 24
for diabetes 25, 75
as favorite food #1 25
heartburn and 109, 112
iron absorption and 23
Cold keys, for nosebleed 190–191
Cold sores 55–56
Colds 48–55, 63, 66, 188
Colitis 56–59
Concord grape juice 31, 64, 125, 133
Constipation
associated with irritable bowel syndrome 61, 144, 145, 147
caused by hypothyroidism 202
from coconut 58, 79
in diverticulitis 84
home remedies 59–63, 82
Corn Huskers Lotion 219
Cornmeal, for nail fungus 179
Cornstarch, for rosacea 15–16
Coughs 53, 63–66, 133
cough syrup recipe 65
ginger remedy 188
Coumadin. *See* Warfarin
Coumarin 73, 74, 200
Crohn's disease 56–58, 79, 145

Curcumin
anti-inflammatory properties 40, 154, 197, 198
anticancer potential 11, 198
antiviral properties 225
for arthritis 11, 152, 154, 198
blood glucose and 76
blood pressure and 128
for bursitis 40
for gout 99, 198
for psoriasis 194, 196–197, 198
Curry
as favorite food #23 198
for gout 99
See also Turmeric
Cuts and bruises 43, 66–69, 90

D

Dandruff 69–72, 179, 195
Dandruff shampoo
antifungal ingredients 70, 179
for nail fungus 179
for skin fungus 211
DermaSmart clothing 89
Diabetes 11, 25, 42, 72–78
foot problems 177
peripheral neuropathy 186, 190
skin tags 212
Vitamin D deficiency 158
Diabetes Diet 239, 240
Diarrhea 78–83
artificial sweeteners and 16, 63, 80, 81, 203
associated with diverticulitis 84, 85
associated with irritable bowel syndrome 59, 144–147
coconut for 58, 78–79, 145
magnesium and 144, 172, 203
probiotics for 82, 83, 93
Diet, acne and rosacea and 16–17
Dietary Approaches to Stop Hypertension (DASH) diet 123, 125, 231, 232–235, 238
Dietary changes, for incontinence 141
Dietary discretion, for diarrhea 80–81
Dietary triggers, of migraines 102
Dill pickle juice 118, 173
Diverticulitis 83, 84–85
Dizziness 166–169
Dry skin 72, 85–87, 88, 91
Duct tape, for warts 221, 222, 224

E

Ear pulling, for hiccups 118–119
Earl Grey tea, as cause of muscle cramps 170
Eczema 16, 83, 85, 88–93, 194, 226
Epley maneuvers 168
Equal sweetener, for joint pain and arthritis 154
Exercise, for high blood pressure 125

F

Fennel seed
as favorite food #11 96
fennel tea recipes 96, 108
for heartburn and gas 95, 96, 108, 110
Fenugreek, for diabetes 75
Fever blisters. *See* Cold sores
Fiber, for constipation 60–61
Fibromyalgia 93–94, 153, 156
Fish and fish oil
for brain function 30, 32, 131
cardiovascular benefits 131
for diarrhea 81
as favorite food #17 131
for high cholesterol and triglycerides 129, 130
for joint pain and arthritis 129, 131, 154–155
for macular degeneration 131, 165
in Mediterranean diet 235–236
omega-3 fatty acids 77, 131, 136, 154, 178
for plantar fasciitis 193
for psoriasis and eczema 195
Flatulence cushion 97
Flaxseed oil
for constipation 61
for eczema 91, 195
for psoriasis 194–195

G

Garlic
anticancer potential 43, 223
as favorite food #24 223
for high blood pressure 125–126, 223
for warts 222
Gas
from broccoli 110
remedies 95–97, 108, 110
Gatorade, for muscle and leg cramps 171
Gin-soaked raisins 155, 157, 201
Ginger
for colds 53–54
as favorite food #22 188
for heartburn 109
for migraines 102
for nausea 186
for vertigo 168–169

Ginger tea 49, 53, 169, 186
recipes 54
Gluten-free diet
for allergies 20
for migraines 103
for muscle and leg cramps 171–172
Golden raisins 155, 157
Gout 94, 98–100, 128, 152, 153, 198
Grape juice
for coughs 64
as favorite food #18 133
for high blood pressure 64, 123, 125, 133
for high cholesterol 64, 129, 133
for joint pain and arthritis 129, 133, 150–151, 157
Grapeseed extract, for high blood pressure 126
Green beans, for canker sores 45
Green olives, for hiccups 119
Green tea
anticancer potential 41, 43
for asthma 24
cardiovascular benefits 41, 92
as favorite food #3 41
for high blood pressure 126–127
for shingles 208

H

Hair color, walnuts and 136
Headaches and migraines 25, 100–104, 188
Heartburn remedies 47, 96, 104–115, 106
Hemorrhoids 116–117
Hiccups 117–121
High blood pressure 66, 121–128, 142
antihypertensive drugs 64, 122
beets as remedy 11, 121, 123
DASH diet 123, 125, 231, 232–235, 238
grape juice as remedy 64, 123, 125, 133
sodium restriction 118, 127
Vitamin D deficiency and 158
High cholesterol 25, 73, 128–135, 202
High triglycerides 128–135
Honey
for coughs 63
for wound care 227, 228, 229
Honey and vinegar
for constipation 61–62
for joint pain and arthritis 156
Hot and spicy soup, for migraines 103–104
Hot flashes 135–140
Hot peppers
capsaicin content 17, 103, 112, 113
as favorite food #14 112
for heartburn 111, 112, 113
Hot water
for bug bites and stings 33
for plantar warts 221
Hydrogen peroxide, for nail fungus 180, 184
Hypertension. *See* High blood pressure

I

Incontinence 140–142
Insomnia 142–144, 172, 185
Instant glue, for eczema 90
Iodine
for nail fungus 180–181, 184
for warts 224
Iron deficiency 22–23, 201–202
Irritable bowel syndrome 58, 78, 80, 92, 113, 144–148, 196

J

Joint pain 100, 120, 129, 148–158

K

Kegel exercises, for incontinence 141
Keys, cold, for nosebleed 190–191
Kidney stones 83, 159–161, 197, 232
Kiwi, for canker sores 45

L

L-lysine
for canker sores 45, 46
for cold sores 55, 56
LEARN diet 238
Leg cramps. *See* Muscle and leg cramps
Lemonade
"hot lemonade" 62
for kidney stones 160–161
Lice 161–163
Liquid bandage 90, 212–213
Listerine, as remedy for
ant bites 36
blemishes 18
body odor 28
dandruff 69–70, 195
lice 162–163
nail fungus 181–182
psoriasis 195
shingles 208
skin fungus 210–211
stinky feet 215
sunburn 217–218
warts 222, 224
Low-carb diet 231, 238–249
acne and 17
for eczema 90
for heartburn 113–114, 231

weight loss and 226
Low glycemic index diet 226–227, 239
Lutein, for macular degeneration 163, 164, 166

M

Macular degeneration 131, 163–166
Magnesium, as remedy for
constipation 62
headaches and migraines 102
high blood pressure 123, 125
insomnia 143–144
muscle and leg cramps 171, 172, 173
restless leg syndrome 172, 202–203
Magnets, for arthritis 156
Mayonnaise, for lice 163
Meat tenderizer, for bug bites and stings 33–34, 36
Mediterranean diet 26, 231, 235–237
Migraines. *See* Headaches and migraines
Milk of magnesia, as remedy for
acne 18–19
bedsores 229
body odor 28, 211
dandruff 70
seborrheic dermatitis 70, 211–212
Minerals
for high blood pressure 125
for muscle and leg cramps 172
Miracle Whip, for dandruff 71
Motion sickness 109, 143, 166–169, 185, 186, 188
Muira puama 205
Muscle and leg cramps 128, 170–177, 178
Mustard
for diabetes 76
as favorite food #21 178
for muscle and leg cramps 176–177, 178
See also Yellow mustard
Mylanta, for canker sores 46–47

N

Nail fungus 177–184
Nausea 113, 147, 184–187, 188
Nerve pain 149, 186–189
Neti pots 20–21, 209–210
Nettle, stinging 21, 142
Nopal cactus, for diabetes 76–77
Nosebleed 190–191
Noxema, for eczema 90–91
Nuts
arginine content 56
DASH diet 233
health benefits 77, 105, 136
Mediterranean diet 237
See also Almonds; Peanuts; Walnuts

O

Oatmeal, for high cholesterol 130
Olive oil, as remedy for
dry skin 87
vaginal dryness 87, 219, 220
Omega-3 fatty acids
in fish oil 77, 131, 136, 154, 178
in walnuts 77, 105, 136
Onion. *See* Raw onion
Oolong tea
blood sugar and 75, 92
for eczema 91, 92
as favorite food #10 92
for high blood pressure 92, 126–127
Orange juice, for kidney stones 160–161
Ornish diet 226, 238

P

Papain 34, 57, 114
Papaya, for heartburn 114
Paper
drinking through, for hiccups 118
nosebleed remedy 191
Peanuts
blood lipids and 105
for hiccups 120
Pectin. *See* Certo
Peppermint oil
for gas 97
for irritable bowel syndrome 145–146
Periodic limb movement disorder 204
Pickle juice
for hiccups 118
for muscle and leg cramps 173
Pineapple
as favorite food #6 57
See also Bromelain
Pineapple juice
for arthritis 156–157
for hiccups 120
Plant stanol esters, for high cholesterol 132
Plantar fasciitis 153, 192–193
Plantar warts 220, 221, 222, 225
Pomegranate
anticancer potential 42, 43
as favorite food #4 42
Pomegranate juice
for diarrhea 81–82
for high blood pressure 42, 123, 125
for joint pain and arthritis 150, 151
Potato, for treatment of warts 224

Power pudding, for constipation 62
Prelief, for canker sores 47
Proanthocyanidins 93, 126
Probiotics, as remedy for
colds and flu 49, 55
diarrhea 82, 83, 93
diverticulitis 85
eczema 91, 93
gas 97–98
irritable bowel syndrome 93, 146–147
Prostate health 21, 41, 42, 142, 198
Psoriasis 16, 69, 128, 148, 193–198
Pycnogenol
for eczema 93
for hot flashes 138–139

Q

Questran, for diarrhea 82
Quinine 170, 172, 174

R

Raisins, gin-soaked 155, 157, 201
Raw onion
for bug bites and stings 34
juice, for burns 37–38
Raynaud's disease 199–200
Recipes
Anti-inflammatory Curcumin Scramble 241
Cardamom Grilled Chicken with Mango Lime Sauce 245–246
Chicken Adobo 51
Cholesterol-Combating Oatmeal 242
Coconut Chicken Soup 51
Coleslaw with Mint 248
Crustless Cauliflower and Red Pepper Quiche 243
Digestive Tea 108
Farmers Market Saag 248–249
Fennel Seed Tea 96
Ginger Pickle 111
Ginger Tea 54
Helen Graedon's Chicken Soup 50
Hot Toddy 54
Joe's Brain-Boosting Smoothie 241–242
Lentil Nut Loaf with Red Pepper Sauce 246–247
Persimmon Punch 111
Pescado al Cilantro 247
Roasted Garlic 249
Simple Salad Dressing 249
Smoothie 134
Sole with Lemongrass Red Bell Pepper Salad 244
Thyme Cough Syrup 65
T's Immuno-tea 200
Red bush tea. *See* Rooibos tea
Red clover salve, for rosacea 19
Red wine
in Mediterranean diet 237, 238
as migraine trigger 101, 102
proanthocyanidins 126
Red yeast rice, for high cholesterol 132, 134
Restless leg syndrome 172, 200–204
Rice diet 123
Rooibos tea, for allergies 21
Rosacea 15–19
Rubbing alcohol
for body odor 28–29
for nail fungus 184

S

Sage, ground, for cuts 69
Saline solution, for sinusitis 209, 210
Salt, blood pressure and 118, 127
Sauerkraut, for canker sores 45, 47
Sea-Bands 143, 185
Sex/libido 205–207
Shingles 149, 207–209
Sinusitis 209–210
Skin fungus 210–212
Skin tags 212–213
Sloe gin, for joint pain and arthritis 157
Soap under bottom sheet
for arthritis 157–158
for leg cramps 173–174
for restless leg syndrome 203–204
Sodium lauryl sulfate 44
Sour cherries, for gout and arthritis 99–100, 152
Soy sauce, for burns 38–39
St. John's wort, for hot flashes 137–138
Stevia sweetener, for high blood pressure 127–128
Stinging nettle
for allergies 21
for incontinence 142
Stinky feet 214–216
Styptic pencils, for canker sores 48
Sugar, for wound care 227, 228–229
Sugar-free gum 61, 63, 80, 81, 107–108
Sugar-free sweets, for constipation 62–63
Sunburn and sun rash 217–218

T

Tea
as heartburn trigger 112
herbal, for insomnia 144
for high blood pressure 125
iron absorption and 23
recipes 54, 96, 108, 200
for stinky feet 215

See also Black tea; Earl Grey tea; Ginger tea; Green tea; Oolong tea; Rooibos tea
Tea tree oil, for nail fungus 183, 184
Testosterone 205–206
Thread, for skin tags 213
Thyme, for coughs 52, 64, 65
Tonic water, for muscle and leg cramps 171, 174
Toothpaste
for bug bites and stings 33, 35–36
sodium lauryl sulfate in 44
Turmeric, as remedy for
arthritis 11, 99, 128, 152, 154, 198
back pain 27
bursitis 40, 128
diabetes 76, 128
digestive problems 115, 196
gout 99, 128, 198
high blood pressure 128
irritable bowel syndrome 148, 196
muscle and leg cramps 128, 175, 177
nerve pain 187, 189
psoriasis 128, 148, 194–197, 198
warts 224–225

U

Udderly Smooth Udder Cream 86, 87
Uric acid, in gout 98, 99, 100, 152, 153
Urinary tract infections 10, 90, 226
Urine, as remedy for
nail fungus 183
stinky feet 216

V

V8 juice, for muscle and leg cramps 175–176
Vaginal dryness 87, 140, 218–220
Vanilla extract, for burns 39
Vertigo 166–169
Vicks VapoRub
for fire ant bites 33, 36
for nail fungus 180
on soles of feet, for coughs 66
Vinegar, as remedy for
athlete's foot 216–217
body odor 29
bug bites and stings 32, 33, 36
dandruff 71–72
diabetes 77–78
dry skin 87
heartburn 114–115
hiccups 120–121
high cholesterol 134–135
muscle and leg cramps 176
nail fungus 184
rosacea 19
warts 225–226
See also Honey and vinegar
Vitamin B12, nerve pain and 187, 189–190
Vitamin C, for sunburn 218
Vitamin D
deficiency 158
for fibromyalgia 94
for joint pain and arthritis 158
for macular degeneration 165
from sun exposure 217
Vitamin E oil, for nail fungus 184
Vitex 206
Vodka, for body odor 29

W

Walnut oil 237
Walnuts
as favorite food #19 136
for high cholesterol 11, 77, 105, 135, 136
omega-3 fatty acids 77, 105, 136
preventive health benefits 11, 105, 136
Warfarin (Coumadin), interaction with
coumarin 73
curcumin/turmeric 40, 128, 154, 197, 198
papain 114
Warts 33, 68, 220–226
Weight loss 226–227
DASH diet 233
with green tea 41
low-carb diet 231
low glycemic index diet 226–227
rice diet 123
Wound care 227–229

Y

Yams, for hot flashes 139–140
Yeast infections 83, 90, 195, 226
Yellow mustard 11, 76, 115, 128, 175–178, 189, 196, 198
Yogurt, as favorite food #9 83
Yogurt, probiotic, as remedy for
colds and flu 49, 55
diarrhea 82, 83
diverticulitis 83, 85
gas 83, 98
irritable bowel syndrome 146–147
Yohimbine 206–207

Z

Zinc
for coughs 66
for macular degeneration 165, 166
for stinky feet 215, 216
Zinc oxide
for body odor 29
in sun blockers 218
Zinc pyrithione 70, 211
Zone Diet 238, 239

THE PEOPLE'S PHARMACY
QUICK & HANDY HOME REMEDIES
Joe and Terry Graedon

PUBLISHED BY THE NATIONAL GEOGRAPHIC SOCIETY

John M. Fahey, Jr., *Chairman of the Board and Chief Executive Officer*

Timothy T. Kelly, *President*

Declan Moore, *Executive Vice President; President, Publishing*

PREPARED BY THE BOOK DIVISION

Barbara Brownell Grogan, *Vice President and Editor in Chief*

Marianne R. Koszorus, *Director of Design*

Carl Mehler, *Director of Maps*

R. Gary Colbert, *Production Director*

Jennifer A. Thornton, *Managing Editor*

Meredith C. Wilcox, *Administrative Director, Illustrations*

STAFF FOR THIS BOOK

Susan Tyler Hitchcock, *Editor*

Barbara H. Seeber, *Text Editor*

Chalkley Calderwood, *Art Director*

Quentin Bacon, *Photographer*

Judith Klein, *Production Editor*

Lisa A. Walker, *Production Manager*

Al Morrow, *Design Assistant*

MANUFACTURING AND QUALITY MANAGEMENT

Christopher A. Liedel, *Chief Financial Officer*

Phillip L. Schlosser, *Senior Vice President*

Chris Brown, *Technical Director*

Nicole Elliott, *Manager*

Rachel Faulise, *Manager*

Robert L. Barr, *Manager*

IMPORTANT NOTE TO READERS

This book is not a substitute for the medical advice or care of a physician or other health care professional. The reader must consult a physician in matters relating to his or her health, especially with regard to any signs or symptoms that may require diagnosis or medical attention. Any health problems that do not get better promptly or get worse should be evaluated by an appropriate clinician and treated properly.

Home remedies are rarely tested in a scientific manner. They should not be considered a substitute for proper medical care. The reader should not assume that the home remedies discussed in this book are safe in all situations. Every treatment has the potential to cause some side effects for some people.

If you suspect that you or someone you care for is experiencing an adverse reaction from an herb, drug, dietary supplement, or home remedy, please consult a knowledgeable health professional immediately.

Throughout this book, the authors make references to specific products whose effects have been reported to them by readers, listeners, and website visitors. In publishing this book, the National Geographic Society does not endorse the use of any specific brands or products.

The National Geographic Society is one of the world's largest nonprofit scientific and educational organizations. Founded in 1888 to "increase and diffuse geographic knowledge," the Society works to inspire people to care about the planet. National Geographic reflects the world through its magazines, television programs, films, music and radio, books, DVDs, maps, exhibitions, live events, school publishing programs, interactive media and merchandise. *National Geographic* magazine, the Society's official journal, published in English and 32 local-language editions, is read by more than 35 million people each month. The National Geographic Channel reaches 320 million households in 34 languages in 166 countries. National Geographic Digital Media receives more than 13 million visitors a month. National Geographic has funded more than 9,200 scientific research, conservation and exploration projects and supports an education program promoting geography literacy. For more information, visit nationalgeographic.com.

For more information, please call 1-800-NGS LINE (647-5463) or write to the following address:

National Geographic Society
1145 17th Street N.W.
Washington, D.C. 20036-4688 U.S.A.

For information about special discounts for bulk purchases, please contact National Geographic Books Special Sales: ngspecsales@ngs.org

For rights or permissions inquiries, please contact National Geographic Books Subsidiary Rights: ngbookrights@ngs.org

Library of Congress Cataloging-in-Publication Data

Graedon, Joe.

The people's pharmacy quick & handy home remedies / Joe Graedon, Terry Graedon.

p. cm.

Index.

ISBN 978-1-4262-0711-2 (pbk.)

1. Herbs--Therapeutic use. 2. Dietary supplements. 3. Alternative medicine. I. Graedon, Teresa, 1947- II. Title. III. Title: People's pharmacy quick and handy home remedies.

RM666.H33G693 2011

615'.321--dc22

2010051751

Printed in the United States of America
11/CW-CML/1